Immunology for Medical Students

IMMUNOLOGY FOR MEDICAL STUDENTS

by

Constantin A. Bona

Mount Sinai School of Medicine
New York

Francisco A. Bonilla

Mount Sinai School of Medicine
New York

harwood academic publishers

chur • london • paris • new york • melbourne

Harwood Academic Publishers

Post Office Box 197
London WC2E 9PX
United Kingdom

58, rue Lhomond
75005 Paris
France

Post Office Box 786
Cooper Station
New York, New York 10276
United States of America

Private Bag 8
Camberwell, Victoria 3124
Australia

Library of Congress Cataloging-in-Publication Data

Bona, Constantin A.
Immunology for medical students/by Constantin A. Bona, Francisco A. Bonilla.
p. cm.
Includes bibliographical references.
Includes index.
ISBN 3-7186-0507-4 (hardcover).—ISBN 3-7186-0508-2 (softcover)
1. Immunology. I. Bonilla, Francisco A., 1960– . II. Title.
[DNLM: 1. Immunity. QW 504 B697i]
QR181.B76 1990
616.07'9—dc20
DNLM/DLC
for Library of Congress 90-4647
CIP

Contents

Acknowledgements

Figure 2.6 and the specimens photographed in Figures 2.2, 4, 11, 14, and 16 were kindly provided by Dr. Edward W. Gresik, associate professor of anatomy at the Mount Sinai School of Medicine of the City University of New York. Figures 2.5, 6.4, 7, 8, 8.4, and 9 were generously given by Dr. Ronald E. Gordon, associate professor of pathology, also of the Mount Sinai School of Medicine of the City University of New York.

Preface

As is apparent from the title of this work, this book is directed primarily at a well-defined group of readers: students of medicine. Medicine is both a science and an art, and in most medical school curricula, these aspects of medicine are presented in that order. Students are first introduced to the voluminous scientific investigations of human physiology and pathology, and are then guided into the hospital proper to become steeped in clinical acumen.

Very few things in life are ever as simple as they are often made to appear, and medical school is no exception. What the medical student most frequently lacks is time. Therefore, an important question often asked is, "Which of the seemingly infinite resources of medical information will tell me what I need to know in the most efficient way?"

We have attempted to compile, in as few words as possible, the fundamental concepts of immunology and some of the latest related research. The information is arranged so that later chapters build on earlier ones, but we have also made reference to chapters containing more extensive discussions of a particular subject when it arises in passing in another context. In addition to describing fundamental immune system physiology, we also introduce many disease entities related to immune system dysfunction and delineate the underlying pathophysiological mechanisms. The references accompanying each chapter contain "classical" investigative works as well as recent monographs and reviews of various topics. We hope that these references will provide interested readers with points of departure from which to further explore this subject.

Although the particular interests of medical students had a role in shaping this book, we hope that any individual having a strong background in biochemistry will find it a useful introduction to immunology. This field is one of the most active areas of research today, and a familiarity with current concepts is important for clinicians in every specialty and for researchers in many biological disciplines. We hope that this volume provides an avenue of ready access to this fascinating area of biology.

Constantin A. Bona
Francisco A. Bonilla

Chapter 1

Introduction

Most of us are aware that our bodies contain something called an "immune system" which protects us from infectious microbes. The notion of immunity as protection from harmful organisms remains the most prominent feature of current immunological paradigms. We have learned recently, however, that the immune system has many more roles than that of a "guardian" at points of entry to the body. The immune system acts to destroy our own cells that develop aberrantly (cancers) as well as to eliminate dead or senescent cells and defunct proteins from the blood and other tissues. Doubtless, the immune system has other equally important physiological roles of which we are not yet aware. We have also discovered that defects or dysregulation of the immune system may result in a tremendous variety of pathologies. Since the immune system interacts with every other system of the body, a knowledge of its fundamental physiology and pathology is essential to every medical specialty.

Immunology arose from early studies of microbiology. Soon after the development of the germ theory of infectious disease in the nineteenth century, investigators discovered that animals possessed mechanisms capable of some degree of defense against disease-causing agents. The word "immunity" is derived from the Latin *immunis* and *immunitas* denoting exemption from military service or taxation. The word was borrowed by early immunologists at the turn of the century to denote the "exemption" from infection that occurred under certain circumstances.

The phenomenon of "acquired immunity" has been known from antiquity in China and Greece. Individuals who survived one plague were observed to remain relatively healthy in subsequent visitations of plague. The Chinese developed the practice of *variolation*. Healthy people were inoculated nasally with powdered crusts from the pustules of individuals stricken by the smallpox (variola) virus. Although a number of people treated in this way probably died as a result, those who survived were protected from future smallpox infection. Variolation evolved into *inoculation*. In this method, a person was scratched over a vein with a needle laden with material from smallpox pustules. This practice spread from India to Turkey where in 1717 Lady Mary Wortley Montagu happened to observe it while visiting there. She introduced this procedure to Europe which suffered greatly from widespread smallpox at that time.

Observing that milkmaids who contracted cowpox, a benign disease, were immune to smallpox, the British physician Edward Jenner conducted a daring experiment in the 1790's. He inoculated two boys with pustules from a woman with cowpox, then inoculated them with smallpox several days later. The youths fortunately suffered no ill consequences, and Jenner had established the foundation of modern immunization. The Latin word for cow is *vacca*, and Louis Pasteur dubbed Jenner's technique *vaccination*, and the injected material the *vaccine*.

At the turn of the century, two principal theories were advanced to explain the phenomenon of immunity. One was the *cellular theory* pioneered by Metchnikoff who observed that starfish contained cells which engulfed and destroyed foreign matter. He later demonstrated that higher organisms (rabbits, for example) also possessed cells able to ingest foreign particles such as anthrax bacilli.

Another prominent hypothesis concerning the mechanism of immunity was the *humoral theory* which held that immunity was due not to cells, but to substances dissolved in the blood. Two early proponents of this theory were von Behring and Kitasato who demonstrated the existence of soluble compounds in blood which neutralized the activity of tetanus and diphtheria toxins. As it happened, the humoral theory predominated and shaped immunological thinking until the 1960's. We now know that cellular and humoral mechanisms together generate immunity.

The protective function of the immune system is often divided into two parts, *specific immunity* and *non-specific immunity*. Mechanisms of specific immunity protect us from only one particular microorganism. That is, specific immunity (also called *adaptive immunity*) acquired by infection with smallpox virus, for example, is of no benefit to us when we encounter rabies virus. On the other hand, non-specific immune mechanisms (also called *innate immunity*) operate against a tremendous variety of infectious agents.

The humoral substances mediating specific immunity were called *antibodies* by the prominent immunologist Karl Landsteiner early in this century. This word embodies the concept of a molecular entity (body) directed specifically against (anti) a particular microorganism or a toxin. The symmetrical concept of a molecule which elicits (generates) antibody production is contained in the word *antigen.*

One of the most fascinating properties of the immune system is that it seems to lie dormant in the context of the multitude of potential antigens comprising our bodies, yet reacts vigorously against antigens in our environment. The immune system is *tolerant* to the antigens of the host. This is one of the most fundamental dichotomies of immunology: *the self non-self discrimination.* The immune system is able to differentiate between antigens foreign to the body, and those existing within it. Vigorous immune responses against self components (*autoimmunity*) are harmful to the host. As the reader will later see, the lines between self and non-self grow less distinct as our knowledge of immune system physiology grows. However, this dichotomy still permeates our conceptions of the organization and regulation of the immune system.

Another important property of the immune system is *immunologic memory.* This is central to the phenomenon of acquired immunity described above. Vaccination introduces an antigen to which the immune system responds. When the antigen is encountered subsequently, the immune system responds more rapidly and vigorously than it did the first time. The immune system's first encounter with an antigen is a learning process resulting in the acquisition and storage of information.

The idea that the immune system has evolved exquisitely to protect us from marauding protoplasm seems inevitable, since this is its function which is most evident to our senses. Whether or not other selective principles, of which we are not yet aware, may be equally important in guiding immune system evolution, only time will tell.

Let us now turn our attention to the physiology of the immune system. We will begin our discussion with a description of the types of cells and tissues which comprise it.

Chapter 2

Cells and Organs of the Immune System

CELLS OF THE IMMUNE SYSTEM

The mammalian immune system consists of several organs, tissues and various cell types circulating in the blood and lymph. The cellular elements of blood are red cells (erythrocytes), white cells (leukocytes) and platelets (thrombocytes). Leukocytes are the "circulating" components of the immune system.

All formed elements of blood derive from a common precursor: the *pluripotent stem cell* (Figure 2.1). The differentiation of stem cells to mature cells is called *hemopoiesis* or *myelopoiesis*. Stem cells first arise in the yolk sac during fetal development. In the sixth week of life, they colonize the liver which remains a *hemopoietic* (blood forming) tissue until birth. Gradually, during the end of gestation and the infant life, hemopoietic cells leave the liver and take up residence in the bone marrow in most of the long bones and in the axial skeleton (pelvis, vertebrae, rib cage, calvarium). In adult life, little hemopoietic marrow remains in long bones, and is found only in the axial skeleton. For clinical examination, marrow may be aspirated from the sternum or the iliac crests (Figure 2.2).

Pluripotent stem cells give rise to five principal types or "lineages" of blood cells, the *erythroid* (red cells), the *thrombocytic* (platelets), and the *granulocytic*, *monocytic*, and *lymphoid* (white cells). An important characteristic of stem cells is their capacity for self-renewal. That is, though mitosis, stem cells give rise to additional stem cells. This ensures perpetuation of a population of cells capable of giving rise to more stem cells and blood elements.

Various microenvironmental conditions within the bone marrow, and soluble factors arising in a variety of tissues, determine the differentiative pathway from stem cells to the progenitors of various lineages. For example, *erythropoietin*, a glycoprotein synthesized by cells within the kidney, increases erythrocyte production. *Colony stimulating factors*

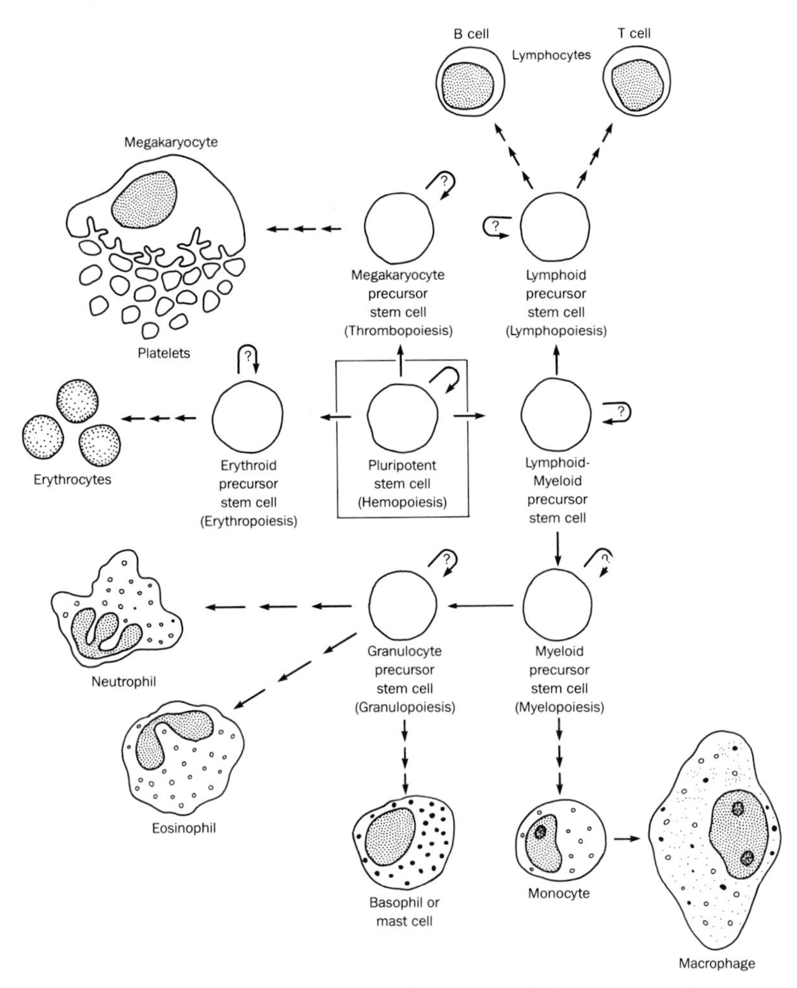

Figure 2.1. *Hemopoiesis.* In the center of the figure is the pluripotent stem cell, the single cell capable of giving rise to all formed elements of blood. The reflexive arrow indicates that this cell may also generate new pluripotent stem cells. Arrayed about the pluripotent stem cell are various intermediate stem cell stages representing committed steps along various pathways of hemopoietic differentiation. The question markes within the reflexive arrows indicate uncertainty concerning the capacity of these cells for self-renewal. The existence of these various intermediate stem cell stages of differentiation is suggested by much research, but they have not been isolated and characterized. These intermediate stem cells proceed through several differentiative steps (not shown) generating mature blood cells. Erythrocytes (in mammals) and platelets are not nucleated cells. Red blood cells extrude their nuclei (which are ingested by macrophages) immediately prior to their exit from the bone marrow. Platelets arise via cytoplasmic fragmentation of the very large megakaryocytes.

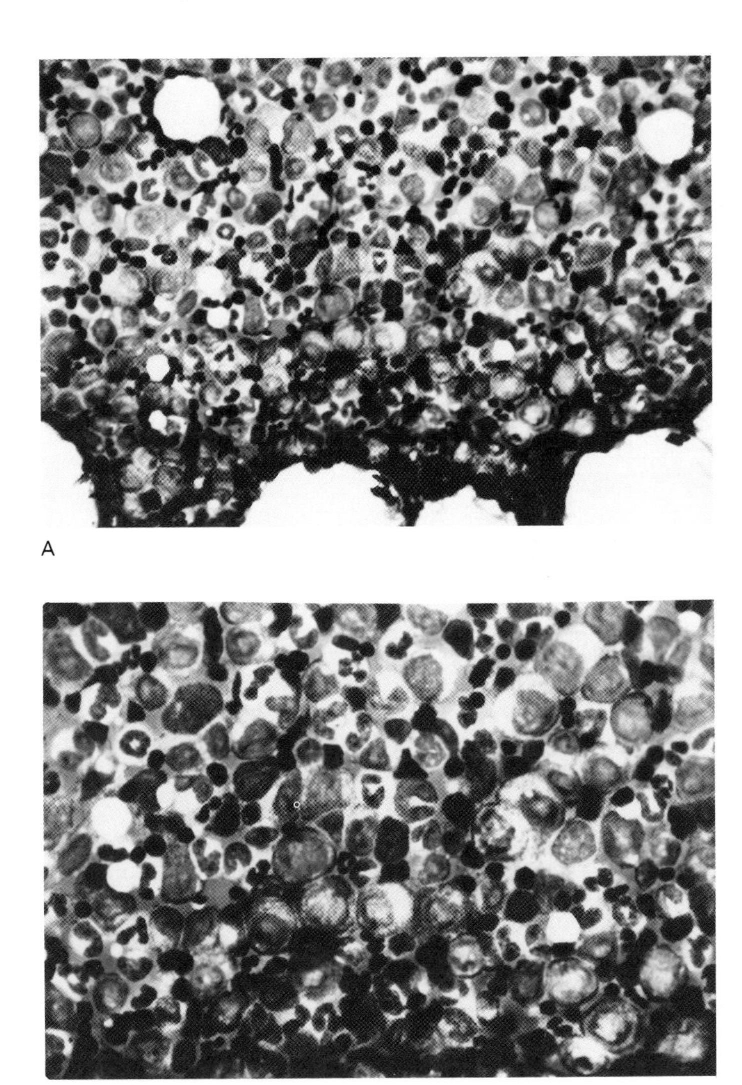

Figure 2.2. *Bone marrow smear.* (Human marrow, Giemsa stain.)

A. *Medium magnification.* Bone marrow may be aspirated with a needle from the sternum, or a core may be obtained from the iliac crest. Aspirated marrow is gently pressed onto a slide for fixing and staining. Since bone marrow is a very loosely organized tissue, almost all structure is lost in this type of preparation. Most of the large cells with large nuclei and basophilic cytoplasm are early erythroid precursors. Many more mature erythrocytes with bright red cytoplasm and small dense nuclei are also present. Many neutrophil precursors are also identifiable. The easiest to recognize are band forms, medium-sized cells with a U-shaped nucleus and pale cytoplasm.

B. *High magnification.* The center of A has been enlarged to show more detail. A few immature eosinophils are clearly visible, one at the center of the figure. (See Color Plate I.)

(*CSFs*), glycoproteins synthesized by some leukocytes and other cell types, may increase production of one or more different lineages. These agents maintain cell division and differentiation, and are additionally pleiotropic since they may also modulate the functions of mature cells. Three different CSFs have been identified. *G-CSF* increases production of neutrophils, and may also stimulate the activity of mature granulocytes. *M-CSF* stimulates mature macrophages and also promotes their differentiation from immature precursors. *GM-CSF* increases production of neutrophils, macrophages and eosinophils, and modulates the activity of mature cells, as well. Many other *cytokines* also regulate hemopoiesis and modulate the activity of mature leukocytes. These substances will be discussed in Chapter 7.

Markers of cellular differentiation

Immunity results from both specific and non-specific protective mechanisms. That is, some immune mechanisms operate against a great variety of pathogens (non-specific) while other mechanisms operate against only one or a few (specific). Broadly speaking, these two types of mechanisms are mediated by different cell lineages, although some overlap in activity exists between distinct cell types.

It is important to keep in mind that not all morphologically similar cells have similar functions, and that similar functions may be carried out by morphologically dissimilar cells. While the synthesis and/or secretion of particular enzymes or the expression of integral membrane glycoproteins often does correlate with morphology, this is not always true. Many methods have been developed to distinguish enzymatic and cell surface differences between microscopically identical cells. Some of these methods and important differences between cell types will be introduced below; others will be saved for later chapters where each cell type will be described in more detail.

A particularly important concept is that of the *cytodifferentiation antigen*. These are proteins or glycoproteins expressed on cell surfaces. Many such antigens have been described, but few have been characterized functionally. They are extremely useful as markers of cell lineages and stages of differentiation. Some cytodifferentiation antigens may be found on several cell types, others on only one. Yet other antigens may even be unique to only one developmental stage of one or more lineages. Any surface molecule, such as a hormone receptor, may serve as a cytodifferentiation antigen, even if we do not specifically call it that.

The utility of a cell-surface protein as a marker depends on its distribution among body tissues. The insulin receptor is not a valuable marker because it is expressed on virtually all cells—it does not serve to distinguish one type of cell from another. The CD3 molecule, on the other hand, is a useful marker since it is only expressed on T cells.

Most cytodifferentiation antigens have been defined with monoclonal antibodies (see Chapter 5) obtained after immunizing animals with various types of cells. These monoclonal antibodies bind to cells expressing the same surface molecules as the cells used for immunization. Since there are hundreds of laboratories engaged in this type of research, it is inevitable that several would independently identify the same cell-surface antigen, but give it different names. A system of nomenclature for these molecules has been established, and is periodically updated as new data become available. The reactions of many monoclonal antibodies with a variety of cell preparations are compared. Antibodies may then be grouped into "clusters", all members of a cluster having similar patterns of reactivity with different types of cells. Thus, each cluster defines one (or a few structurally similar) cytodifferentiation antigen(s). The antigens are named with a "cluster designation", the letters "CD" followed by a number. A partial list of the defined CD antigens is given in Table 2.I.

What techniques do we use to identify cells expressing a particular glycoprotein or enzyme? One very powerful method uses antibodies conjugated to fluorescent dyes, or to enzymes which can generate reaction products visible in a microscope or spectrophotometer. This is called *immunostaining* (see Figures 2.10 and 8.4).

Various fluorescent staining techniques may also be used in conjunction with a *cytofluorograph* which measures the amount of fluorescence emitted by individual cells in a suspension exposed to ultraviolet light. This process is called *cytofluorimetry* or *flow cytometry*. The *fluorescence-activated cell sorter* (*FACS*) is a cytofluorograph having the additional capability of separating the cells in a suspension based on their fluorescence intensity (Figure 2.3).

Effector cells of non-specific immunity

Cells of the granulocytic lineage, *neutrophils*, *eosinophils*, and *basophils*, and the monocytic lineage, *monocytes* and *macrophages*, are effectors of non-specific immunity. The first stage of cellular development committed to generating these lineages is the *myeloid stem cell*

Table 2.I. DISTRIBUTION AND FUNCTION OF SOME DEFINED CLUSTER DESIGNATIONS

Name	Distribution	Function
CD1 (a, b, and c)	thymocytes, dendritic cells, some B cells	?
CD2	all T cells	ligand for CD58
CD3	all T cells	associated with T cell antigen receptor
CD4	$T_{h/i}$ cells	binding to MHC class II
CD5	all T cells, some B cells	?
CD7	all T cells	?
CD8	$T_{c/s}$ cells	binding to MHC class I
CD11a	leukocytes	LFA-1 α chain
CD11b	macrophages, granulocytes, NK cells	CR3 α chain
CD16	macrophages, granulocytes, NK cells	$Fc_{\gamma}RIII$
CD18	same as CD11	β chain with CD11
CD19	B cells	?
CD20	B cells	ion channel?
CD21	some B cells	CR2
CD23	eosinophils, some B cells	$Fc_{\varepsilon}RII$
CD25	activated T and B cells, macrophages	IL-2 receptor β chain
CDw32	myeloid and B cells, platelets	$Fc_{\gamma}RII$
CD35	monocytes, granulocytes, B cells	CR1
CD44	lymphocytes	lymphocyte homing?
CD55	broad	DAF
CD56	NK cells, activated lymphocytes	?
CD58	leukocytes, epithelial cells	ligand for CD2
CD64	monocytes	$Fc_{\gamma}RI$
CD71	proliferating cells	transferrin receptor

Abbreviations: CR1, C3b receptor; CR2, C3d/Epstein-Barr virus receptor; CR3, iC3b receptor; DAF, decay accelerating factor; $Fc_{\gamma}RI$–RIII; IgG Fc receptors; $Fc_{\varepsilon}RII$, low-affinity receptor for IgE Fc; LFA-1, lymphocyte function-associated antigen 1; MHC, major histocompatibility complex; NK, natural killer cell; $T_{c/s}$, cytotoxic and suppressor T cells; $T_{h/i}$, helper/inducer T cells. Adapted from Knapp et al., 1989.

(Figure 2.1). An earlier stage capable of giving rise to erythrocytes, granulocytes, monocytes and megakaryocytes has been described, but its precise relationship to pluripotent stem cells and more differentiated types has not been established. The name "granulocyte" describes the abundance of small cytoplasmic granules (membrane-bound vesicles) in these cells. The differentiation of granulocytes is called *granulopoiesis.*

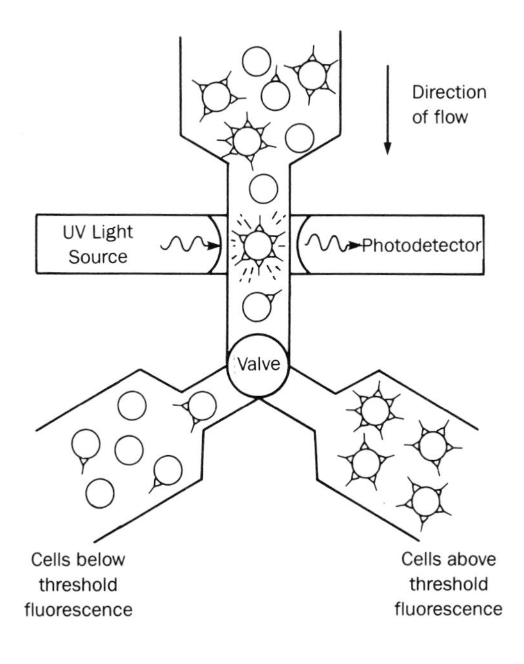

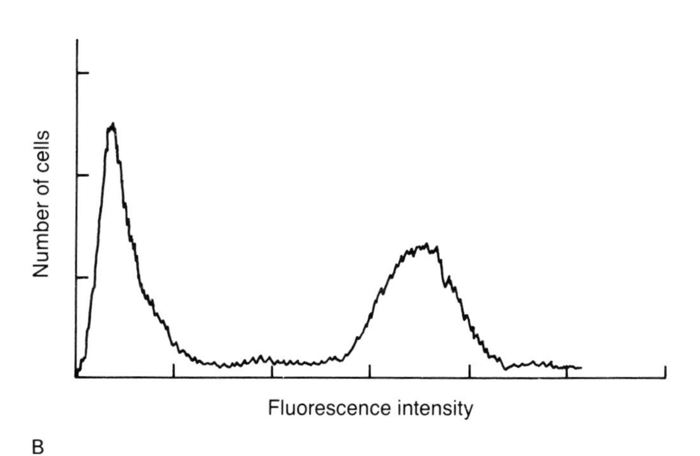

Figure 2.3. *See next page for caption.*

Figure 2.3. *Fluorescence-activated cell sorter* (*FACS*).

A. *Principle of operation.* A suspension of cells treated with a fluorescent reagent (the probe) is dispersed into droplets which pass between an excitatory ultraviolet light source and a photodetector measuring light emitted from the fluorescent probe. The "valve" is not mechanical, but an electric field orienting the motion of the droplets, which bear electrostatic charges. Droplets fluorescing above a threshold value are separated from those that do not.

B. *Sample output of a cytofluorograph.* The data are displayed as a plot of the number of cells exhibiting various fluorescence intensities. When the area under the curve is concentrated near the origin, most of the cells bound little of the probe. Increased binding of the probe shifts the curve to the right. This plot shows that the suspension contains two populations of cells, one does not bind the probe, the other does.

C. *Two-dimensional cytofluorimetry.* A cytofluorograph is capable of simultaneously analyzing the fluorescence of two probes having different excitation and emission spectra. The results of this analysis are displayed with the fluorescence intensity of each probe on perpendicular axes, the numbers of cells indicated by contour lines. This representation is analogous to that of a mountain peak on a counter map. In the example shown, three populations of cells are identified: one binds neither probe, another exhibits marker 1 but not 2, the third displays both markers.

Neutrophils

Neutrophils (also known as *polymorphonuclear leukocytes* or *PMNs*) are the most numerous of the white blood cells, comprising 60–70% of the total number at a concentration of 3–6,000/mm^3 of blood. These cells are 10 to 12 μ in diameter with abundant, slightly basophilic cytoplasm, and a heterochromatic nucleus (Figures 2.4A and 2.5). Nuclear morphology distinguishes two subsets of neutrophils: immature cells called *band cells* (or *band forms*) with an indented (kidney-shaped) nucleus, and mature cells called *segmented cells* in which the nucleus has two or three separate lobes joined by thin chromatin filaments.

Neutrophil granules are 0.3–0.8 μ in diameter and stain weakly with basic dyes (cationic in solution) such as hematoxylin or methylene blue, and acidic (anionic) dyes such as eosin. When neutrophils are stained with a mixture of these dyes, two types of granules are distinguished. The *azurophilic* or *primary granules* have a pale purple color, while the *specific* or *secondary granules* are barely visible in this type of preparation. Both of these types of granules are different forms of lysosomes. The azurophilic granules are the larger of the two (0.5–0.8 μ), and appear dense in electron micrographs. Specific granules are smaller (0.3–0.4 μ in size). Another type of neutrophil granule is the *peroxisome.* This structure contains enzymes which reduce oxygen and hydrogen peroxide. The neutrophil cytoplasm also contains a large Golgi apparatus, mitochondria, some rough endoplasmic reticulum, and small and large polyribosomes. Additional types of neutrophil granules have been identified, but they have not yet been well characterized.

The half-life of a neutrophil after leaving the bone marrow is quite short, on the order of four to ten hours. This cell plays a prominent role in the defense against microbes which replicate extracellularly, predominantly bacteria and fungi.

Eosinophils

Eosinophils are 9–12 μ in diameter, and represent about 3% of all leukocytes (Figures 2–4B and 2–6). These cells have a bilobed or indented nucleus usually located at the cell's periphery. The eosinophil's most prominent feature is an abundance of positively charged granules which stain bright pink or red with eosin, which gives these cells their name. Eosinophils are also easily identified in electron micrographs where their granules are seen to contain a unique crystalloid core structure. These are crystals of the enzyme *lysophospholipase.* The granules also contain *eosinophil cationic protein* (*ECP*) and *eosinophil major basic protein* (*MBP*). Both of these proteins are able to kill bacteria (e.g., *S. aureus* and *E. coli*) *in vitro.* They may act

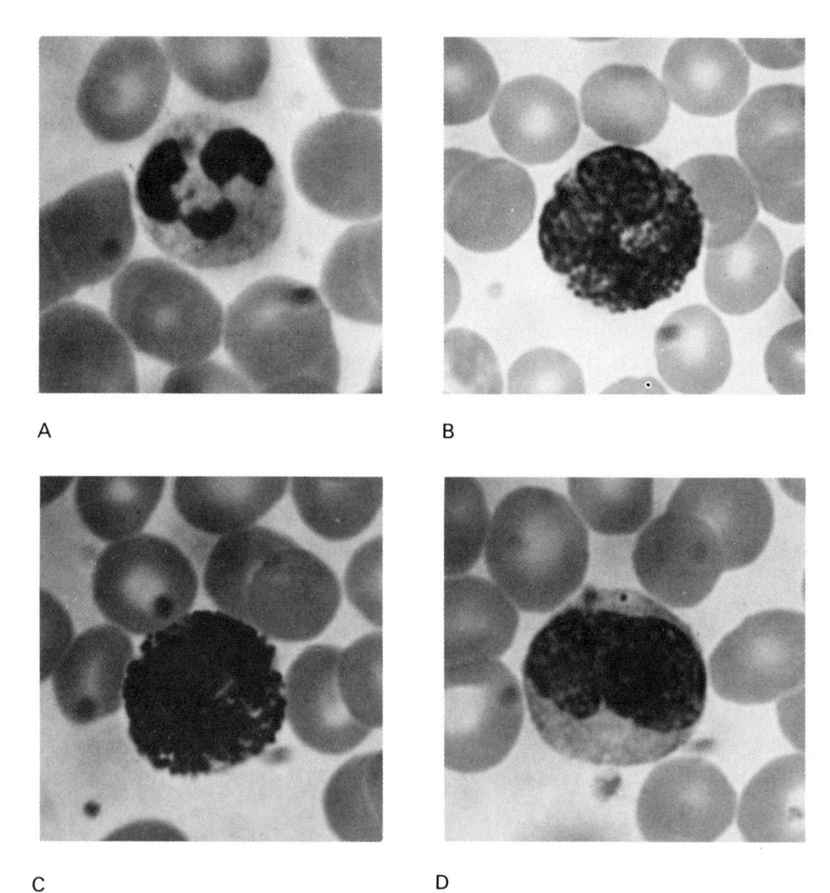
A
B
C
D

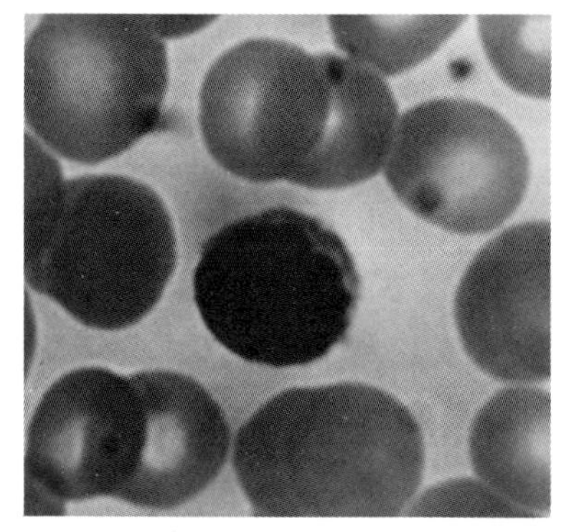
E

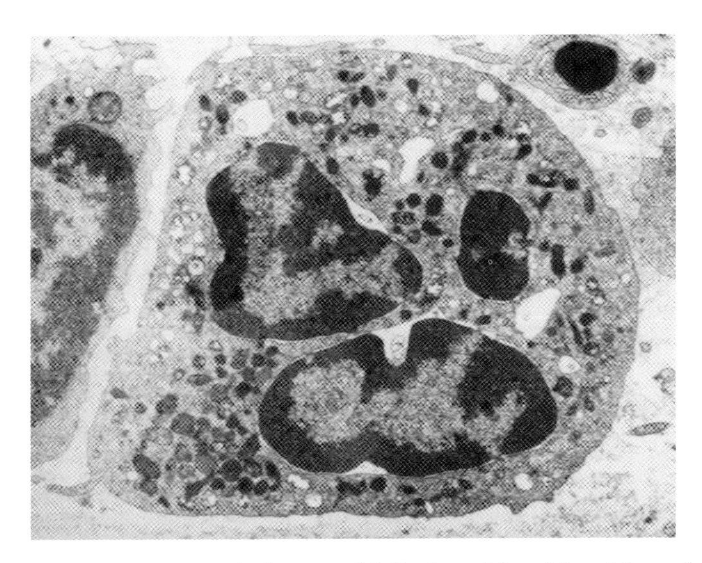

Figure 2.5. *Electron micrograph of a neutrophil.* Portions of three lobes of the nucleus of a human neutrophil are visible in this section. The cytoplasm is packed with a variety of granules. Phagocytic vacuoles are mostly clear, azurophil granules are the darkest. Specific granules are lighter and may be difficult to distinguish from the cytoplasm. Granules of intermediate density may be phagolysosomes or other types of granules.

◀

Figure 2.4. *The five principal types of leukocytes.* (Human peripheral blood smear, Giemsa stain.)

A. *Neutrophil.* This cell has the characteristic dense nucleus with multiple lobes (three here) joined by thin chromatin filaments. The abundant cytoplasm contains many weakly-staining granules.

B. *Eosinophil.* This leukocyte also has a segmented nucleus (two lobes here), and much cytoplasm densely packed with eosinophilic granules.

C. *Basophil.* The nucleus of these cells has irregular morphology, and is less segmented and less dense than neutrophil or eosinophil nuclei. The nucleus is often obscured by the many darkly staining basophilic granules filling the cytoplasm.

D. *Monocyte.* These large cells have a plump irregular or kidney-shaped nucleus, and a weakly basophilic cytoplasm. The cytoplasmic staining is due primarily to the azurophil granules.

E. *Lymphocyte.* This, the smallest leukocyte, consists almost entirely of a round dense nucleus, with only a thin rim of basophilic cytoplasm. (See Color Plate II.).

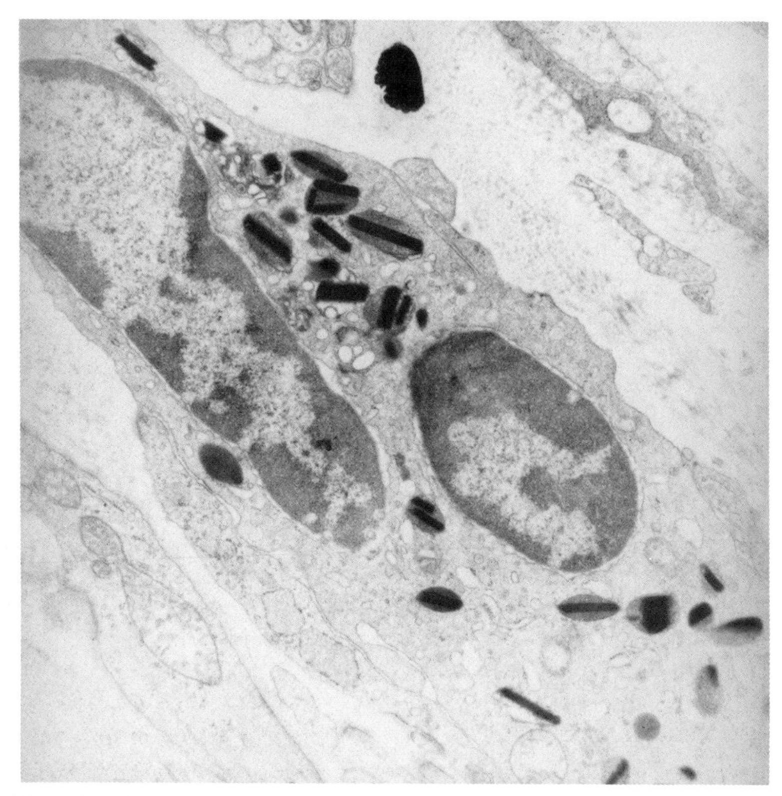

Figure 2.6. *Electron micrograph of an eosinophil.* These cells are unmistakable in electron micrographs. The cytoplasm of this rabbit eosinophil contains many granules with one or more dense rod-shaped crystals of lysophospholipase.

by altering the permeability of the bacterial cell membrane. Granule contents may be released when the cell is activated. Granules may also fuse with phagocytic vacuoles, exposing ingested microbes to their toxic effects. Eosinophils bear receptors for the constant region of IgG antibody, a low-affinity receptor for the constant region of IgE antibody, and a receptor for the C3b fragment of complement. The significance of these receptors will be described in later chapters.

Eosinophils appear to have important roles in handling complexes of antibodies and antigens, controlling (suppressing) inflammation, and protecting against infestation by higher-order eukaryotic parasites such as helminths. An increased number of circulating eosinophils (*eosinophilia*) is a relatively non-specific indicator of pathology and can be seen in a wide variety of disease states.

Basophils

Basophilic leukocytes comprise less than 1% of the total white blood cell population. They are 8–10 μ in diameter, with a large round nucleus often obscured by many ovoid granules (0.3–0.8 μ in size, Figure 2.4C). These cells were named for the intense staining of their granules with basic dyes. The granules are rich in arylsulfatase, β glucuronidase, histamine and heparin.

Basophils enter a variety of tissues such as the lamina propria underlying respiratory and gastrointestinal mucosae, subdermal tissues, mesenteric lymph nodes, and the peritoneal cavity. Basophils residing in tissues other than blood are known as *mast cells* (Figure 2.7). Mast cells are a heterogeneous group, two subsets having been defined by differences in chemical staining and enzyme content. These are the "connective tissue" mast cells, and the "mucosal" mast cells. The functional significance of these distinctions is not yet completely understood.

Mast cells bear high-affinity receptors for the constant region of IgE antibody. These cells play a prominent role in inflammation and immediate hypersensitivity (allergy, see Chapter 10).

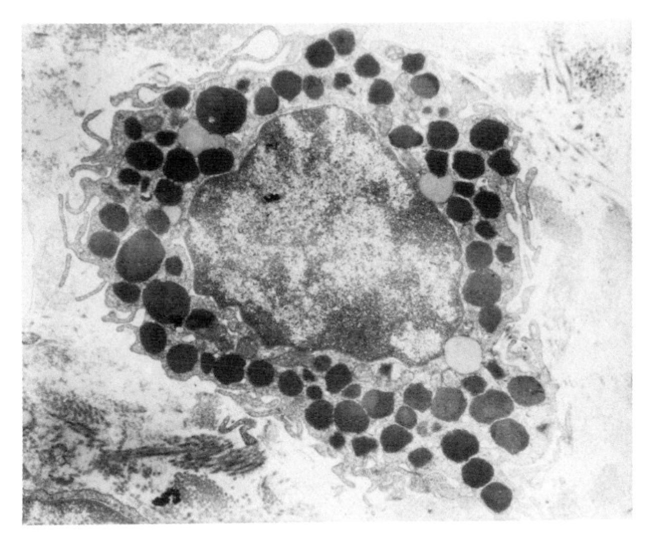

Figure 2.7. *Electron micrograph of a mast cell.* Basophils and mast cells are relatively easy to identify in electron micrographs. This human mast cell from perirectal connective tissue is typical with its cytoplasm crowded with large electron-dense granules.

Monocytes and macrophages

After leaving the bone marrow, monocytes (Figure 2.4D) soon leave the circulation and become "fixed" in other tissues, or enter internal cavities such as the peritoneal, pleural, or pericardial spaces. Upon entering these areas, monocytes may undergo morphological change and are then referred to by a variety of names, depending on the tissue in which they reside. The most common designation is macrophage. Monocyte-derived cells are a subset of the *mononuclear phagocyte system* (*MPS*, Table 2.II), also called the *reticuloendothelial system* (RES). These cells all have one feature in common: they avidly ingest (phagocytose) finely particulate substances. The origins of all cell types within the MPS has not been determined. It appears that blood monocytes give rise at least to macrophages in connective tissue and body cavities, alveolar macrophages and Kupffer cells, as well as those accumulating in inflammatory reactions. Tissue macrophages are long-lived, surviving months or years.

Blood monocytes are 9–15 μ in diameter. Their nuclei may be round, oval, or indented, and usually contain one or two prominent nucleoli. The abundant cytoplasm has a finely granular texture due to the large number of lysosomes and mitochondria. Newly-formed macrophage lysosomes are called *immature granules*. Completely developed lysosomes are the *azurophil granules*. A large Golgi apparatus and much rough endoplasmic reticulum are also observed. Macrophages are often much larger than monocytes, as much as 20–25 μ in diameter.

Several enzymes are useful markers of monocytes and macrophages. *Peroxidase*, which converts hydrogen peroxide to water and oxygen, is detected by the black pigment formed when it acts in the presence of diaminobenzidine and hydrogen peroxide. *Lysozyme*, an enzyme which

Table 2.II. CELLS OF THE MONONUCLEAR PHAGOCYTE SYSTEM

Name	Location
Alveolar macrophages	Lung
Chondroclast	Cartilage
Endothelial cells	Blood and lymph vessels
Histiocytes	Connective tissues
Kupffer cells	Liver
Langerhans cells	Skin
Macrophages	Lymphoid tissue
Mesangial cells	Glomerulus
Microglia	Brain
Monocytes	Blood
Osteoclast	Bone

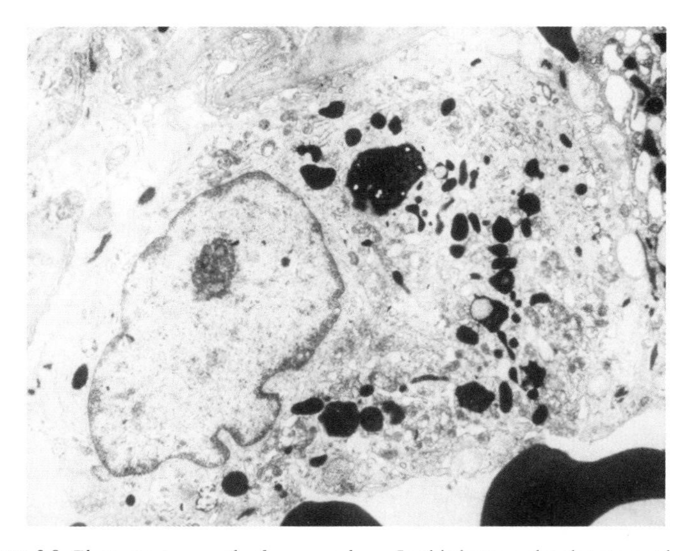

Figure 2.8. *Electron micrograph of a macrophage.* In this human alveolar macrophage we see a prominent nucleolus, scattered mitochondria and rough endoplasmic reticulum, and many dense granules of several shapes and sizes. Small dense vesicles may be azurophil granules, immature granules are less dense. The larger inhomogeneous granules are phagolysosomes which probably contain ingested particulate air pollutants.

degrades components of bacterial cell walls, is also abundant in macrophages.

Macrophages (Figure 2.8) are important in inflammation and in defense against extracellular pathogens. In inflammatory lesions they may fuse together giving rise to *multinucleate giant cells.* These are huge cells (up to 50 μ in diameter or more) in which the nuclei of all the cells which joined together are frequently visible. Two types of giant cells may be found. In the *Langhan's giant cell* the nuclei are arrayed at the periphery of the cell. In the *foreign body giant cell* the nuclei are scattered in the cytoplasm, and engulfed particles may also be visible.

Effector cells of specific immunity

The cells mediating specific immune responses are called *lymphocytes.* These are small and round, 7–9 μ in diameter, with a large nucleus and a very thin rim of cytoplasm containing few lysosomes and mitochondria, but numerous ribosomes (Figures 2.4E and 2.9). Lymphocytes are grouped into functionally distinct, but morphologically identical subsets.

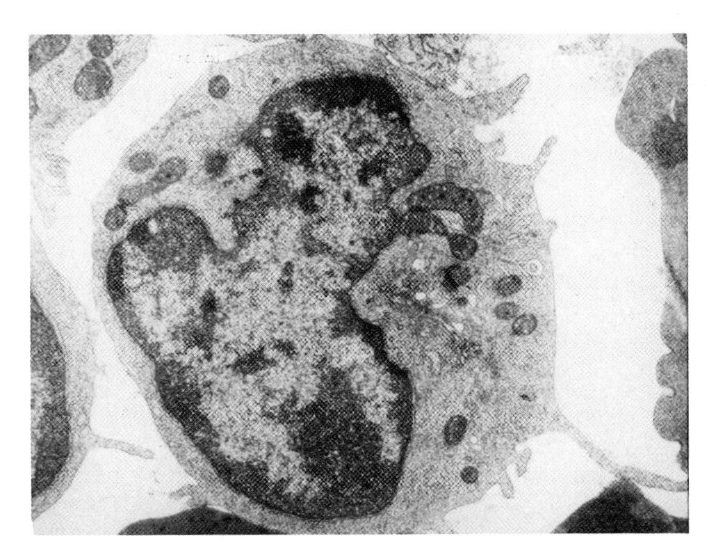

Figure 2.9. *Electron micrograph of a lymphocyte.* This is a typical human peripheral blood lymphocyte. The large nucleus comprises most of the cell. The only prominent organelles in the cytoplasm are mitochondria.

The first subdivision defines two classes: *B lymphocytes* and *T lymphocytes.* Birds possess a lymphoid organ associated with the intestine near the cloaca called the *bursa of Fabricius.* Cells developing in this organ were designated B (bursal) cells. Mammals do not have an equivalent organ, and B cells develop in the liver during fetal life, and in the bone marrow after birth. T cells are so-named because they mature in the thymus.

B lymphocytes

The major distinguishing characteristic of mature B cells is synthesis of antibodies and their presence on the cell surface. Each cell bears approximately 10^5 such molecules. Thus, B cells are most easily identified by substances reacting with surface immunoglobulin. The reagents most widely used for this purpose are antibodies which react with other antibodies. These "anti-immunoglobulins" may be conjugated with fluorescent dyes or enzymes and used in a variety of ways to detect cells with surface or cytoplasmic antibodies (Figure 2.10). B lymphocytes express several cytodifferentiation antigens which will be described in Chapter 6. Antibody production by B cells constitutes the *humoral immune response.*

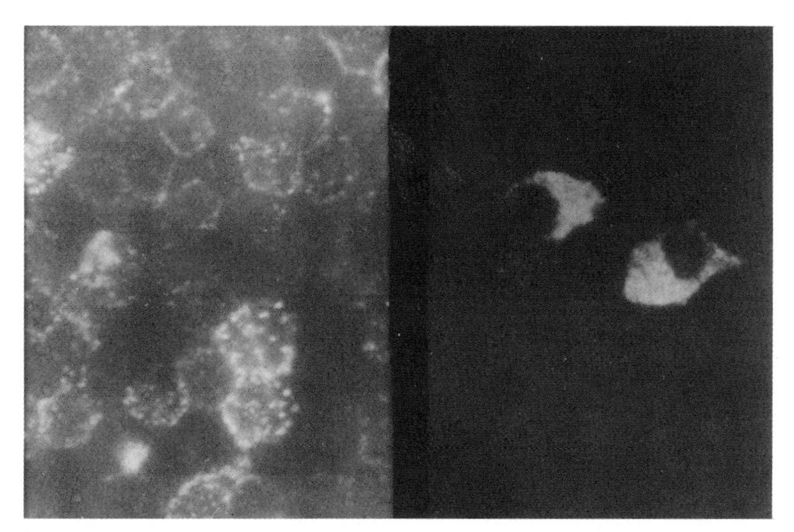

Figure 2.10. *Immunofluorescence staining.* The probe used in these micrographs is an anti-idiotypic antibody (see Chapter 5) labelled with the fluorescent dye rhodamine. The antibody binds to immunoglobulin on the surface of unfixed B lymphocytes (A), and in the cytoplasm of fixed plasma cells (B).

T lymphocytes

Immature cells destined to become T lymphocytes (called *prethymic lymphocytes*, or *pre-T-cells*) migrate from the bone marrow to the thymus where they complete their development. Mature T cells leave the thymus and enter lymphoid tissues.

Mature T cells are a very heterogenous group of cells with two main subsets: effector T cells (*cytotoxic* and *delayed type hypersensitivity*) which are responsible for the selective killing of cells; and regulatory T cells (*helper/inducer*, and *suppressor*) which modulate their own activity and that of other cells. T cells express several CD antigens which will be described in Chapter 7. T cells mediate the *cellular immune response*.

Other lymphoid cells

Another subpopulation of lymphoid cells are called *natural killer (NK) cells*. Morphologically, these cells are *large granular lymphocytes (LGL)*, and may be distinguished from B and T cells by their functions and expression of CD antigens. Natural killer cells were so-named because they were discovered in unimmunized (naive) mice. These cells are able to destroy many different kinds of cells, even from different species, but exhibit considerably greater effect against tumor cells. NK cells may have an important role in the body's defense against some viral infections, and neoplastic cells.

NK cells are capable of destroying their target cells by two mechanisms. One requires antibodies bound to the target cell's surface, the other does not. The process in which antibodies form a bridge between cytotoxic cells and their targets is known as *antibody-dependent cell-mediated cytotoxicity* (*ADCC*). Granulocytes and some macrophage subsets can also perform this function. Cytotoxicity will be discussed in Chapter 7.

Diseases of aberrant hemopoiesis

Several genetically-determined diseases affect stem cell differentiation and lead to congenital absence of one or more of the five principal lineages of blood cells. Several syndromes are grouped under the name *severe combined immunodeficiency disease* (*SCID*). These conditions are characterized by absence or defective function of one or more subsets of cells mediating specific immunity, with variable deficits of non-specific immunity as well. Afflicted infants are extremely susceptible to infection, and generally do not survive beyond the first year of life.

Both X-linked and autosomal recessive modes of inheritance are observed in SCID syndromes, about 75% of patients are male. All of these immunodeficiencies appear to arise from defects at the stem cell level. Although the lymphoid stem cell has not been isolated, its existence is strongly suggested by a form of SCID in which there is complete lack of lymphocytes with normal numbers of erythrocytes, platelets, granulocytes, and monocytes. Maternal antibodies (acquired by the fetus through the placenta, and by the newborn in breast milk) are variably protective for the first six months of life, after which infections with viruses, bacteria, and fungi are inevitable. Vaccines containing attenuated live viruses, such as the oral polio vaccine, can cause fatal infections in infants with SCID.

Adenosine deaminase deficiency is a SCID syndrome with autosomal recessive inheritance. This enzyme converts the purine nucleotides adenosine and deoxyadenosine to inosine and deoxyinosine, respectively. Without this enzyme, intracellular adenosine concentrations are much higher than normal. This is toxic to lymphocytes which die soon after they differentiate. Deficiency of the enzyme *purine nucleoside phosphorylase* is another form of SCID with autosomal recessive inheritance.

At present, the single therapy which may be curative for SCID is bone marrow transplantation. This procedure provides a source of stem cells which can give rise to lymphocytes and other cell types and reconstitute immune function. The most common complication of this procedure is the *graft-versus-host reaction* in which the transplanted bone marrow cells generate an immune response against the recipient (see Chapter 8).

LYMPHOID ORGANS

Lymphoid tissues and lymphatic vessels constitute the *lymphatic system.* The *central lymphoid organs* are those areas in which lymphocytes develop from stem cells into mature *immunocompetent* cells. In humans, these organs are the bone marrow and the thymus. Mature lymphocytes are distributed throughout the circulatory and lymphatic systems and occur in high concentrations in certain tissues and organs. These are the *peripheral lymphoid organs*: the *lymph nodes, spleen,* and lymphoid tissue distributed in the mucosa of the upper respiratory tract (*bronchial-associated lymphoid tissue,* or *BALT*), and the intestines (*gut-associated lymphoid tissue,* or *GALT*). BALT and GALT are often grouped together under the acronym, *MALT, mucosal-associated lymphoid tissue.* Recently, the importance of lymphoid tissue associated with the skin has been appreciated, giving us the acronym *SALT* (*skin-associated lymphoid tissue*).

Central lymphoid organs

Bone marrow

Red bone marrow (Figure 2.2), where hemopoiesis occurs, is a loosely organized tissue. A fine meshwork of *reticular fibers* holds *primitive reticular cells,* masses of dividing and developing blood cells, and occasional large adipocytes. In addition to creating the bone marrow architecture, reticular cells probably play an active role in hemopoiesis.

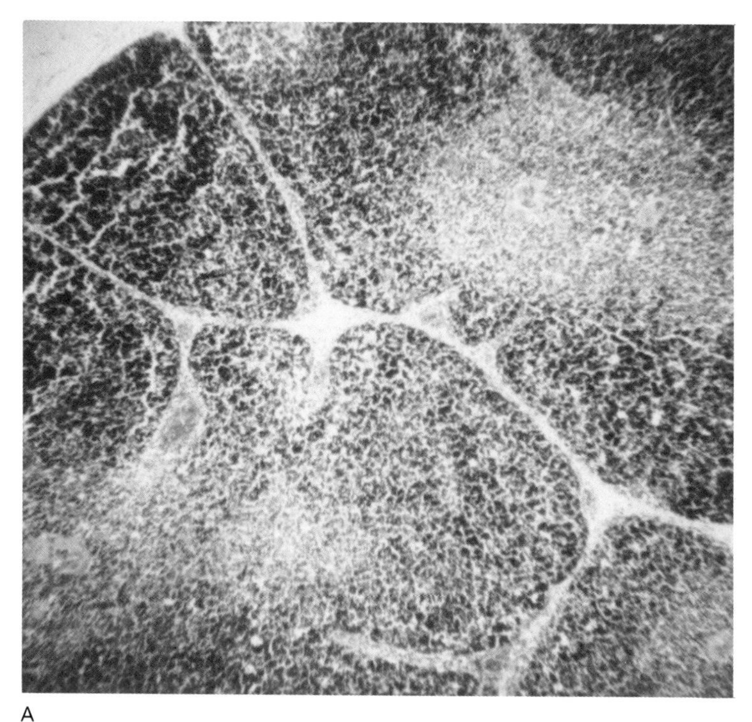

A

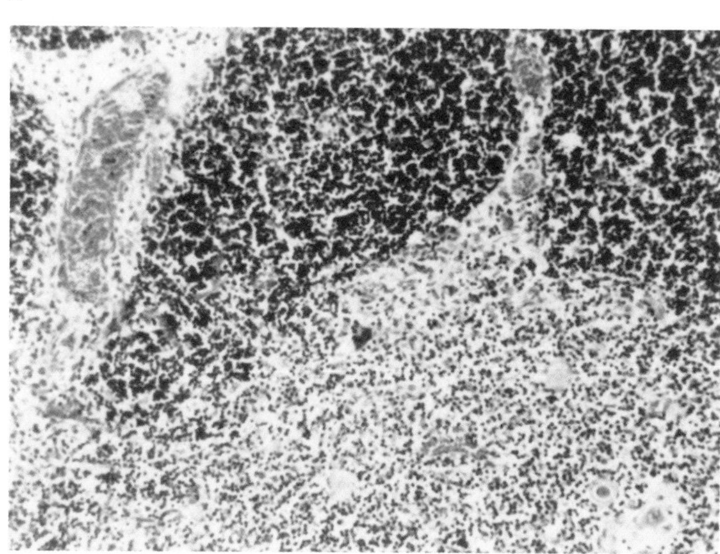

B

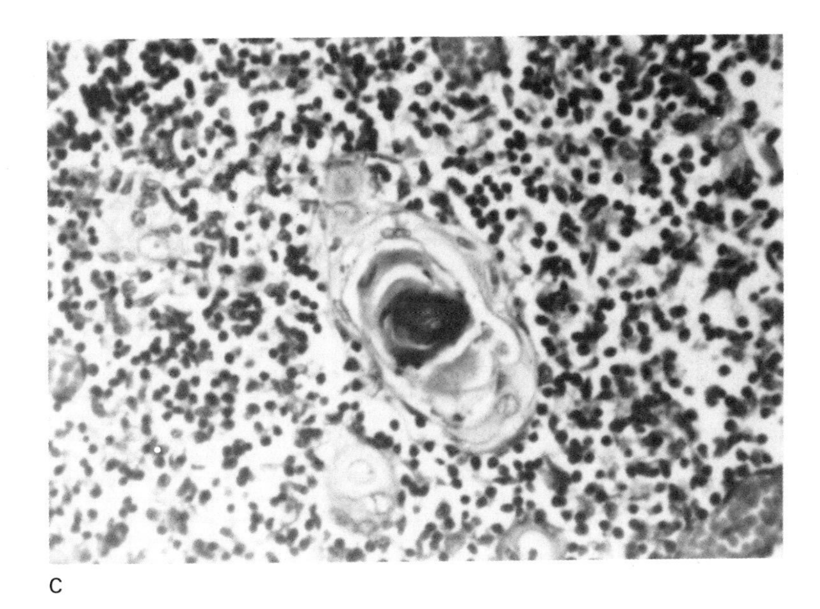

C

Figure 2.11. *The Thymus.* (Human neonate, hematoxylin and eosin stain.)

A. *Low magnification.* Connective tissue septa, traversed by large blood vessels, divide the thymus into lobules. The very dense peripheral cortex is clearly distinguished from the more central, less dense medulla.

B. *Medium magnification.* The cortex (top of micrograph) is packed with small lymphocytes, they are more diffuse in the medulla (bottom). Large cells with lightly-staining cytoplasm and pale nuclei may be found in both areas, but they are easier to see in the medulla. These are epithelial cells.

C. *High magnification.* This structure is a *Hassal's corpuscle*, an aggregate of epithelial cells. Cells in the center are probably dead, and may even be keratinized. These structures are found only in the thymus. (See Color Plate III.)

They may influence dividing cells either by secreting trophic factors, by direct cell-cell contacts, or both. Little is known of the precise role of reticular cells in hemopoiesis. Newly formed blood elements enter the circulation through the walls of large *venous sinuses.*

The thymus

The thymus has two *lobes* surrounded by a thin *capsule* of connective tissue extending into the lobe and dividing the parenchyma into *lobules* (Figures 2.11 and 2.12). The peripheral zone of each lobule is called the *cortex*, while the more central area is the *medulla.*

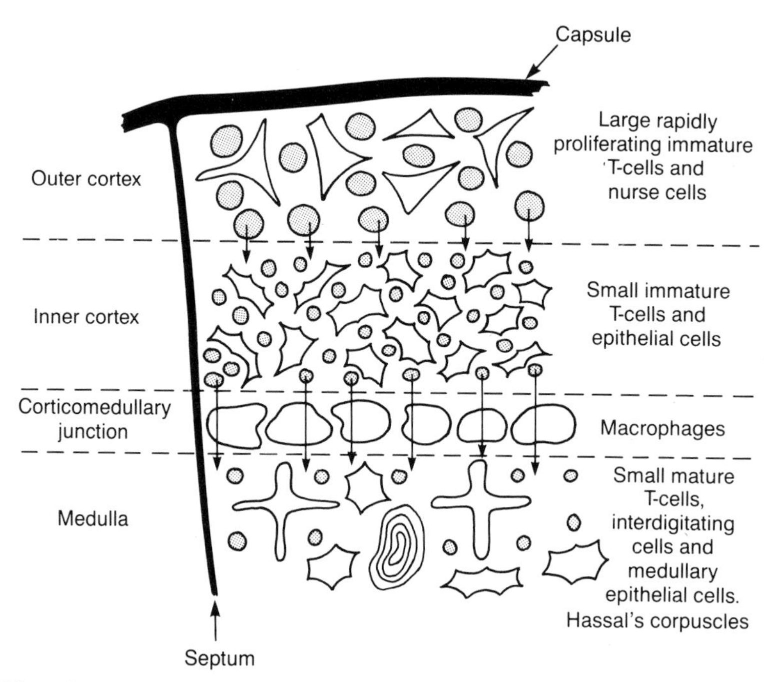

Figure 2.12. *Diagram of a thymic lobule.* Pre T cells enter the outer cortex where they become large rapidly dividing cells. Here they interact with specialized epithelial cells called *nurse cells.* Thymocytes migrate toward the medulla, become smaller and divide less rapidly. They then traverse the corticomedullary junction which contains a high density of macrophages. It has been speculated that these macrophages are responsible for the death of the majority of thymocytes. Additional types of specialized epithelial cells populate the thymic medulla along with mature thymocytes and Hassal's corpuscles. Mature T cells enter the circulation through venules in the medulla and the corticomedullary junction.

The cortex contains the majority of thymic lymphocytes, some macrophages, and various specialized epithelial cells. The latter play an important part in thymocyte development, but their role is only beginning to be understood. Several distinct types have been identified: *nurse cells, cortical* or *medullary dendritic cells* and *reticular cells.* Functional distinctions between these cell types are still obscure. The progeny of rapidly dividing lymphocytes in the outer cortex migrate toward the medulla. At the boundary of cortex and medulla (*the corticomedullary junction*), the majority of these lymphocytes degenerate and die. Some may be ingested by macrophages. This process is believed to be the destruction of those T cells which either cannot function properly, or which could produce harmful reactions against self components if allowed to enter the circulation. The details of the interactions of thymocytes with the various types of specialized thymic epithelial cells still need much clarification. We will return to our discussion of intrathymic T cell development in Chapter 6.

The medulla contains only 3–5% of the total number of thymic lymphocytes. This region also contains epithelial cells. Macrophages are rare here, granulocytes somewhat less so. The most distinctive histological feature of the thymic medulla are *Hassal's corpuscles*. These structures are large (often > 100 μ) aggregates of concentrically arrayed epithelioid cells which stain bright pink with eosin. The cells in the center of these bodies often degenerate. Hassal's corpuscles may be sites and/or products of the death and degradation of thymic lymphocytes. Mature thymocytes enter the circulation through the walls of *postcapillary venules* at the boundary between cortex and medulla.

The thymus develops in the embryo as an epithelial outgrowth of the third and fourth pharyngeal pouches. In the sixth week of gestation, primitive mesenchymal and neural crest cells seed the epithelial structure which then becomes a site of T cell lymphopoiesis. Development of the parathyroid glands (important regulators of blood calcium and phosphate levels) parallels formation of the thymus. The clinical manifestations of an immunodeficiency disease known as *congenital thymic aplasia* (or the *DiGeorge syndrome*) are readily explained by this conjoint development. In this disease, the third and fourth pharyngeal pouches fail to develop, leading to congenital absence of the thymus and parathyroid glands. This results in deficient cell-mediated immunity and hypocalcemic tetany. Thymus transplants have been successful in restoring immune function in infants who survive tetany.

The thymus is an active lymphopoietic organ throughout childhood. During adolescence the thymus involutes. Lymphocyte numbers gradually decrease as they are replaced by adipose tissue, but they do not disappear entirely. Experimental lymphocyte depletion leads to reactivation of lymphopoiesis in roden thymuses. The thymus may involute abnormally (*accidental involution*) as the result of severe emotional or physical stress (disease, radiation, nutritional deficiency, etc.).

Peripheral lymphoid organs

After maturing in central lymphoid organs, lymphocytes enter the circulation and peripheral lymphoid organs and tissues. Lymphocytes, having travelled through various parts of the body, congregate in these areas and interact with one another and with granulocytes and macrophages, and specialized epithelial cells (reticular cells) which form the tissue stroma. These cellular interactions constitute an exchange of information about conditions encountered in various parts of the body. This exchange occurs via soluble factors and direct cellular contacts involving specialized cell surface glycoproteins and receptors. Some lymphocytes may remain for relatively long periods of time (days) in one location, but the majority constantly recirculate in blood and lymph, never staying in one place for more than a few hours. Incessant remixing of the 10^{12} cells comprising the human immune system is crucial for the efficient interaction of different cell types, and for ensuring that cells capable of warding off microbial invasion arrive at the point of entry in a timely manner (Figure 2.13).

Lymphocyte migration is not just a random mixing of cells in various areas of the body. Lymphocytes bear cell surface molecules such as the CD44 family of glycoproteins which are receptors for molecules expressed on endothelial cells in particular lymphoid tissues. One type

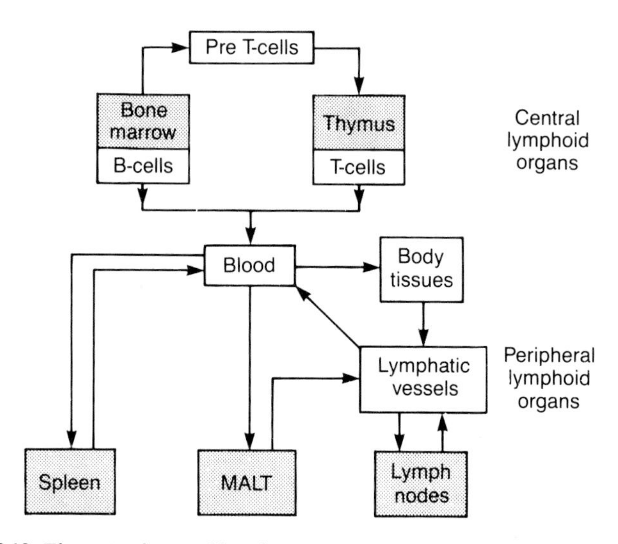

Figure 2.13. *The recirculation of lymphocytes.* Lymphocytes mature in central lymphoid organs and then enter the blood. From here they circulate through the spleen, *MALT* (*mucosal-associated lymphoid tissue*), lymph nodes and other body tissues such as the skin. Lymphocytes may reside in a particular locus for a few hours or days, perhaps becoming activated and proliferating, perhaps not. Eventually, the system of lymphatic vessels returns cells to the blood to begin the journey again.

of receptor may direct lymphocytes specifically to the gut-associated lymphoid tissue, while another might direct migration to the respiratory tract, etc.

The spleen

The spleen is a fist-sized organ situated left-most in the upper abdominal cavity underlying ribs 9, 10, and 11. In addition to the immunological functions carried out by the many lymphocytes resident there, the spleen is a site of removal of senescent blood cells and serum proteins, and particulate substances from the blood. Macrophages carry out this "housekeeping" function.

A connective tissue capsule encloses the splenic parenchyma which consists of two morphologically distinct tissues called *red pulp* and *white pulp* (Figures 2.14 and 2.15). The white pulp contains lymphoid cells aggregating in structures called *lymphoid follicles* or *nodules* (see below), and in *periarteriolar lymphoid sheaths.* The follicles contain mostly B cells, and are called *B-dependent areas,* whereas the periarteriolar sheaths contain mostly T cells (*T dependent areas*). The pulp surrounding periarteriolar sheaths is called the *marginal zone,* an area where arterioles end and dump their contents into the loose matrix of splenic tissue.

The spleen is not essential for life. Because of its very soft internal consistency, the spleen cannot be repaired when it ruptures traumatically, and is usually removed when this occurs. Individuals with sickle cell anemia are often functionally asplenic due to multiple infarctions during sickling crises. Asplenism is associated with somewhat increased susceptibility to infection with encapsulated bacteria (e.g., pneumococci). For this reason, asplenic individuals are commonly immunized with pneumococcal capsular polysaccharides.

Lymph nodes

Lymph nodes (Figures 2.16 and 2.17) are lymph filters situated along lymphatic vessels. Lymph, leukocytes, microbes and particles enter nodes through *afferent lymphatics* and flow through a system of *lymph sinuses* containing a fine meshwork of reticular cells with scattered macrophages. The periphery of the node is the *cortex*, the center is the *medulla.* Leukocytes may also enter the node through cortical venules. Lymphocytes are densely packed in the cortex, they are less crowded

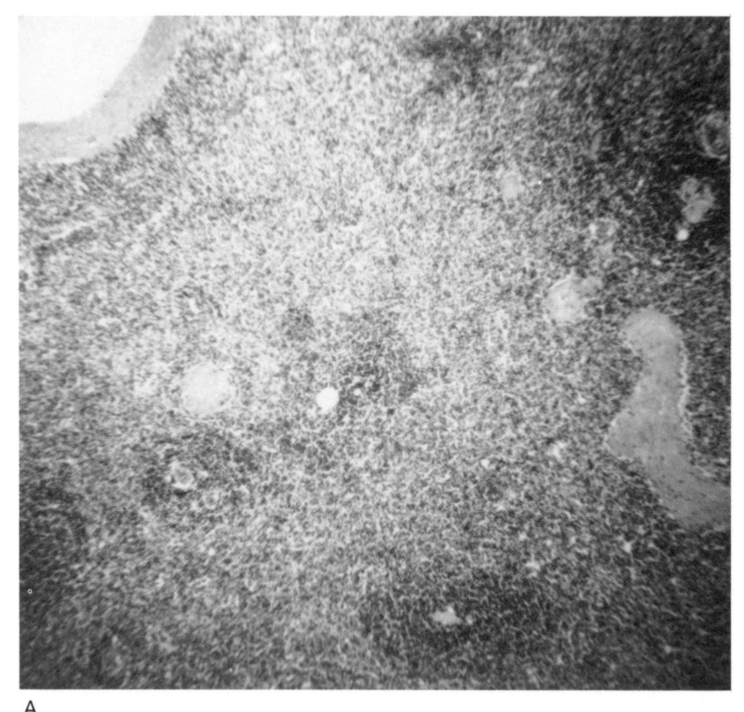

A

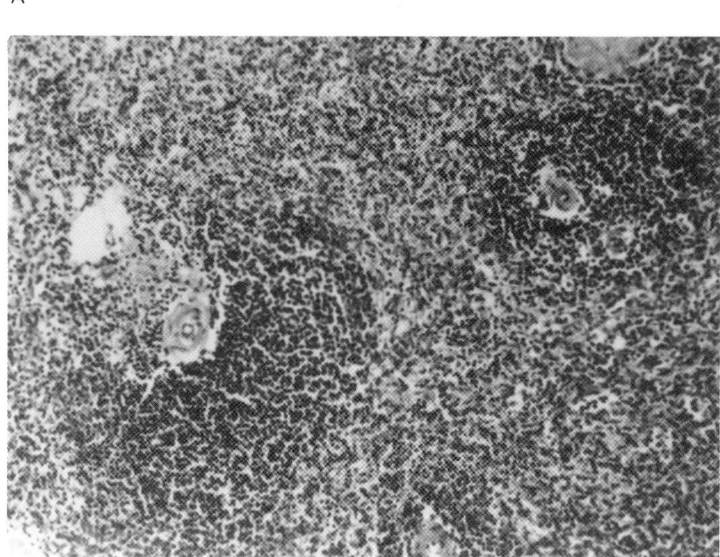

B

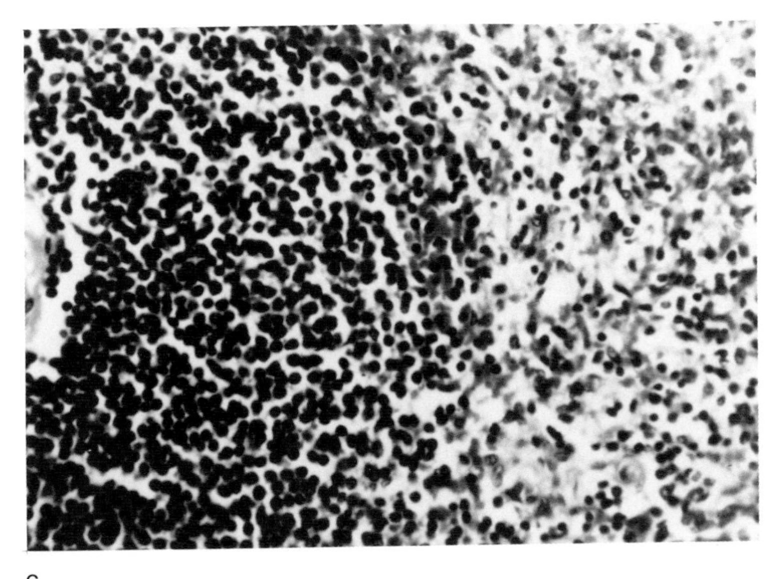

C

Figure 2.14. *The spleen.* (Adult human, hematoxylin and eosin stain.)

A. *Low magnification.* The thick connective tissue capsule (upper left) is continuous with trabeculae projecting into and partially segmenting the spleen. The majority of splenic tissue is red pulp, scattered areas of white pulp (dense and basophilic) are seen.

B. *Medium magnification.* The periarteriolar lymphoid sheaths of white pulp appear as islands in a sea of red pulp containing sinusoids filled with erythrocytes. Sinusoids are separated from one another by pulp cords formed mainly of reticular cells and reticular fibers. Many macrophages are found in the pulp cords as well.

C. *High magnification.* The white pulp (left) is crowded with small lymphocytes. Many erythrocytes and reticular cells (and some lymphocytes) may be seen in the red pulp (right). (The reader should note that this specimen contains a slightly higher proportion of lymphocytes than are normally found in the red pulp. This individual must have had some pathology causing lymphocyte accumulation in the spleen.) (See Color Plate IV.)

in the medulla where the parenchyma is organized into *medullary cords* interdigitating with the *medullary sinuses.* Macrophages are numerous in the medulla. The lymph node parenchyma is enveloped by a connective tissue capsule continuous with *trabeculae* partially segmenting the node. Cells and lymph exit the node through *efferent lymphatics* and eventually reenter the blood.

Cortical lymphocytes aggregate into nodules, also called follicles. Two types are defined: *primary follicles* have a uniform cellular density; *secondary follicles,* also called *germinal centers,* have a dense outer zone with a less dense interior. B cells are found in the follicles and the outer

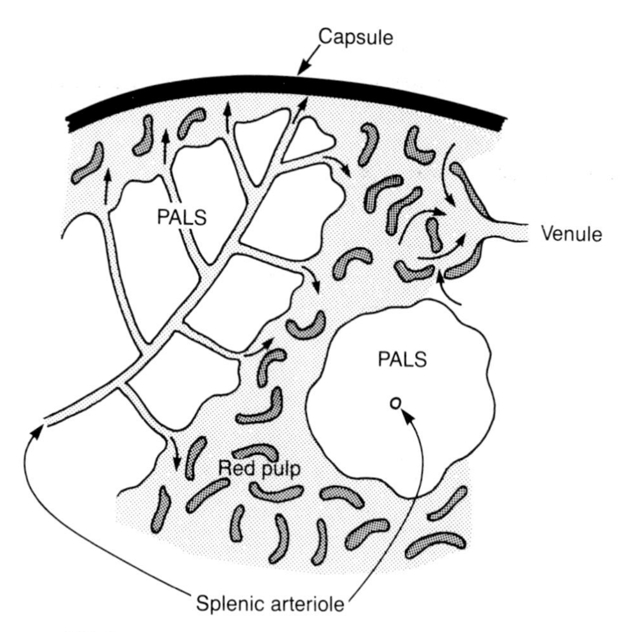

Figure 2.15. *Diagram of the spleen.* The red pulp is a very loosely organized tissue, a collection of large sinusoids with a sparse stroma of reticular cells and large numbers of macrophages. The *periarteriolar lymphoid sheath* (*PALS*, *white pulp*) is similar to the cortex of lymph nodes and occasionally contain lymphoid follicles (see Figure 2.16). For clarity, the PALS is shown here well-demarcated from the red pulp. It is evident from the micrographs in Figure 2.14, however, that the transition between these two types of splenic tissue is not so abrupt.

cortex. T cells are principally located at the boundary of cortex and medulla, although some T cells are found in the B cell areas.

The lymphoid follicles are regions of lymphocyte activation. Cells proliferate in the cortex, some migrate into the medulla, others exit in efferent lymph. As B cells are activated, they become antibody-secreting *plasma cells.* When present, these cells are found scattered in the cortex and in higher numbers in the medulla. Antibodies are secreted into the lymph.

Musocal-associated lymphoid tissue (MALT)

Lymphocytes also aggregate to form follicles in the lamina propria underlying the basement membranes of the mucosal epithelia of the

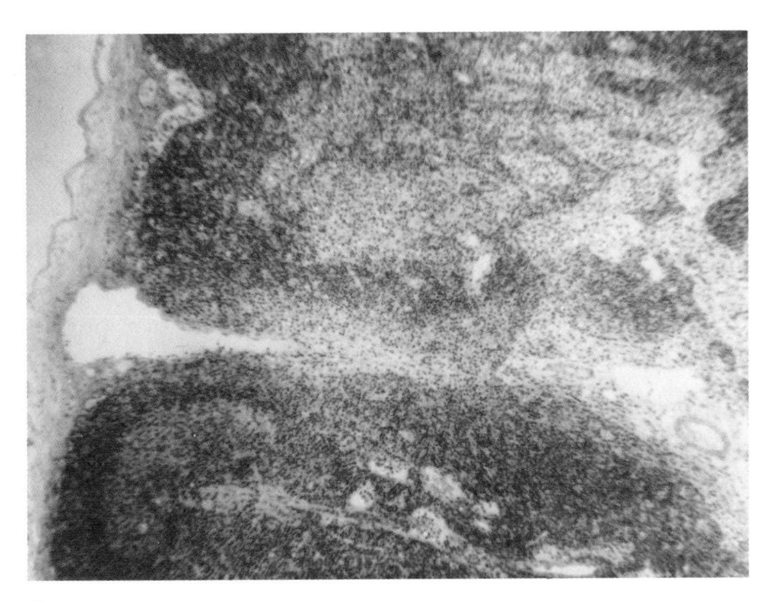

A

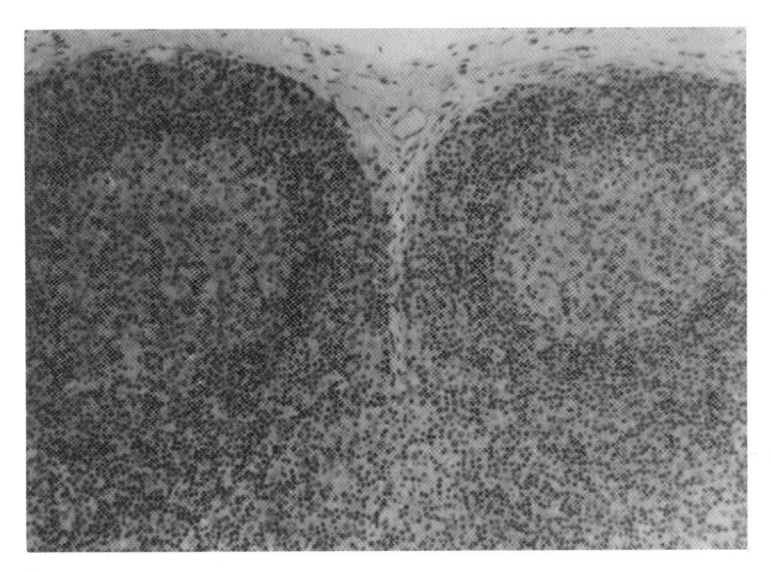

B

Figure 2.16. *See next page for caption.*

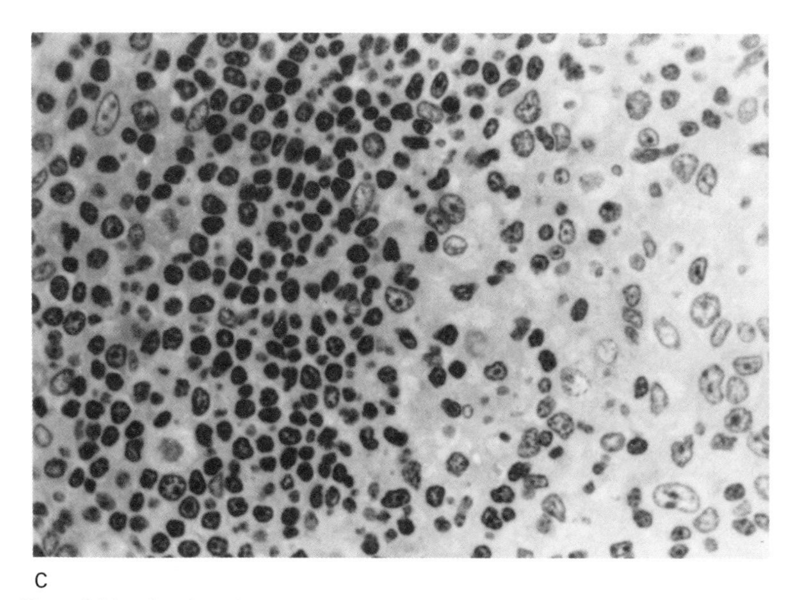

Figure 2.16. *A lymph node.* (Monkey, hematoxylin and eosin stain.)

A. *Low magnification.* Left-most in the figure is the capsule, a portion of the system of sinuses is visible underneath it. The dense cortex at the left blends into the medulla at the right. Medullary sinuses (light) and medullary cords (dark) may be distinguished.

B. *Medium magnification.* Two germinal centers at the periphery of the lymph node cortex.

C. *High magnification.* Small lymphocytes and reticular cells form the capsule of the germinal center (left). The lighter central area (right) contains fewer lymphocytes, many dendritic cells and scattered macrophages. (See Color Plate V.)

gastrointestinal and respiratory systems. In the pharynx, lymphoid follicles are associated with epithelial invaginations to form the *tonsils*, and large numbers of follicles are also found in the walls of the distal ileum where they form structures known as *Peyer's patches.* Lymphocytes are diffusely spread under respiratory epithelia in the trachea and bronchi.

Skin-associated lymphoid tissue (SALT)

Lymphoid cells are also present in the skin. A functionally distinct subpopulation of T cells may home to skin tissue. The skin also contains *dendritic cells.* Several types had been distinguished: *Langerhans' cells, intermediate cells, interdigitating cells,* and *veiled cells.* These cells express class II histocompatibility antigens, cell surface glycoproteins important

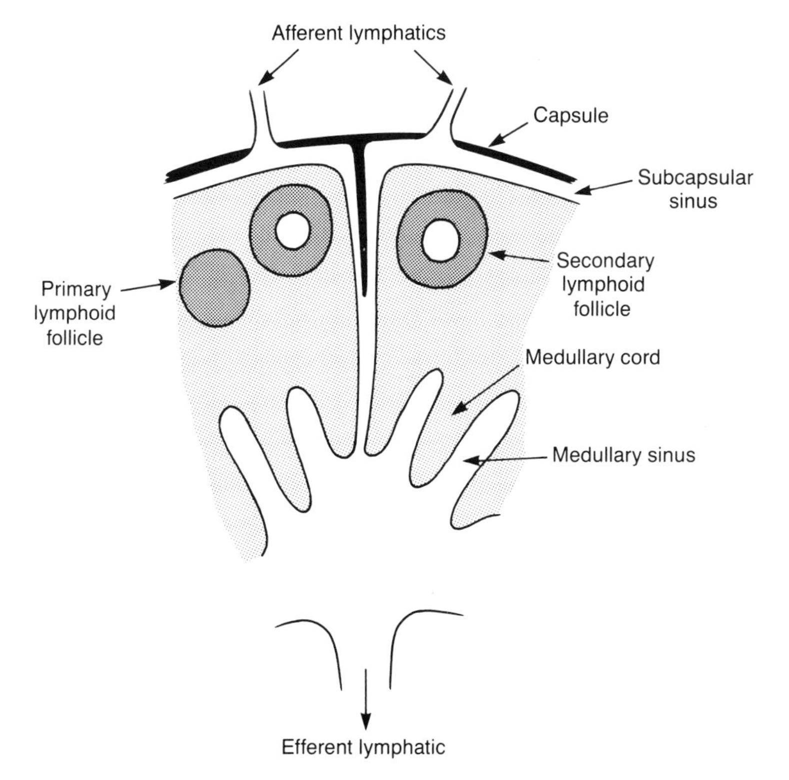

Figure 2.17. *Diagram of a lymph node.* Afferent lymphatic vessels empty into the subcapsular sinus. The system of sinuses follows connective tissue trabeculae and becomes convoluted with the *medullary cords* in the center of the node. Microbes and other foreign materials enter through afferent lymphatics or cortical venules (not shown). Lymphocytes may be activated and proliferate in cortical germinal centers. Activated and recirculating lymphocytes as well as their secreted products (e.g., antibodies and cytokines) exit nodes in the efferent lymph.

in the interactions of lymphocytes with each other and with other types of leukocytes and other cells (see Chapter 8). These cells have important functions in immune responses originating in the skin. One such function is antigen presentation (see Chapter 7). Dermal tissue also contains many mast cells and macrophages. Keratinocytes produce several factors modulating the activities of skin lymphoid cells, including G-, M-, and GM-CSFs, as well as several cytokines (see Chapter 7). Thus, in addition to providing a physical barrier, the skin is an area of immune surveillance and reaction.

Having briefly introduced the cellular elements of the immune system and their development and distribution in the body, we next consider some important aspects of the many diverse molecules which may enter our bodies, and interact with our immune systems.

SOURCES AND SUGGESTED ADDITIONAL READING

Amman, A. J. & Hong, R. (1989) Disorders of the T-cell system. In E. R. Stiehm, ed., *Immunologic Disorders in Infants and Children*, Third Edition, W. B. Saunders Company, Philadelphia, pp. 257–315.

Andreeff, M. & Welte, K. (1989) Hematopoietic colony-stimulating factors. *Semin. Oncol.*, **16**:211–299.

Athanasou, N. A., Hall, P. A., d'Ardenne, A. J., Quin, J. & McGee, J. (1988) A monoclonal antibody (anti-L-35) which reacts with human osteoclasts and cells of the mononuclear phagocyte system. *Br. J. Exp. Pathol.*, **69**:309–319.

Bos, J. D. & Kapsenberg, M. L. (1986) The skin immune system. Its cellular constituents and their interactions. *Immunol. Today*, **7**:235–240.

Brandtzaeg, P. (1988) Immunobarriers of the mucosa of the upper respiratory and digestive pathways. *Acta Otolaryngol. (Stockh.)*, **105**:172–180.

Good, R.A. (1987) Bone marrow transplantation symposium: bone marrow transplantation for immunodeficiency diseases. *Am. J. Med. Sci.*, **294**:68–74.

Kamarck, M. E. (1987) Fluorescence-activated cell sorting of hybrid and transfected cells. *Methods Enzymol.*, **151**:150–165.

Knapp, W., Dörken, B., Rieber, P., Schmidt, R. E., Stein, H. & von dem Borne, A. E. G. Kr. (1989) CD Antigens 1989. *Blood*, **74**:1448–1450.

Kupper, T. S., Horowitz, M., Birchall, N., Mizutani, H., Coleman, D., McGuire, J., Flood, P., Dower, S. & Lee, F. (1988) Hematopoietic, lymphopoietic, and proinflammatory cytokines produced by human and murine keratinocytes. *Ann. N. Y. Acad. Sci.*, **548**:262–270.

Lehrer, R. I., Szklarek, D., Barton, A., Ganz, T., Hamann, K. J. & Gleich, G. J. (1989) Antibacterial properties of eosinophil major basic protein and eosinophil cationic protein. *J. Immunol.*, **142**:4428–4434.

Müller, W., Peter, H. H., Wilken, M., Jüppner, H., Kallfelz, H. C., Krohn, H. P., Miller, K. & Rieger, C. H. L. (1988) The DiGeorge syndrome. I. Clinical evaluation and course of partial and complete forms of the syndrome. *Eur. J. Pediatr.*, **147**:496–502.

Nakamura, H. & Ayer-Le Liévre, C. (1986) Neural crest and thymic myoid cells. *Curr. Top. Dev. Biol.*, **20**:111–115.

Nieuwenhuis, P. & Opstelten, D. (1984) Functional anatomy of germinal centers. *Am. J. Anat.*, **170**:421–435.

Powell, R. W., Blaylock, W. E., Hoff, C. J. & Chartrand, S. A. (1988) The protective effect of pneumococcal vaccination following partial splenectomy. *J. Surg. Res.*, **45**:56–59.

Shanahan, F., MacNiven, I. A., Dyck, N., Denburg, J. A., Bienenstock, J. & Befus, A. D. (1987) Human lung mast cells: distributions and abundance of histologically distinct subpopulations. *Int. Arch. Allergy Appl. Immunol.*, **83**:329–331.

Shau, H., Gray, J. D. & Mitchell, M. S. (1988) Studies on the relationship of human natural killer and lymphokine-activated killer cells with lysosomal staining and analysis of surface marker phenotypes. *Cell. Immunol.*, **115**: 13–23.

Steinmann, G. G. (1986) Changes in the human thymus during aging. *Curr. Top. Pathol.*, **75**: 43–88.

Stoolman, L.M. (1989) Adhesion molecules controlling lymphocyte migration. *Cell*, **56**: 907–910.

Torok-Storb, B. (1988) Cellular interactions. *Blood*, **72**: 373–385.

Chapter 3

Antigens

Antigenicity, immunogenicity, and tolerogenicity

Antigens are substances interacting with specific immunoglobulins on the surface of or secreted by B cells, and with T cell antigen receptors. Antigens are often conceptualized in rather gross terms. We speak of immune responses to proteins or complex polysaccharides, and envision the protein or sugar (antigen) as a single unit, a macromolecule of protein or sugar. It is most instructive for the immunologist, however, to view these large molecules in the same way the immune system does. A, B or T cell interacting with a complex macromolecule does not contact all parts of it, their receptors only recognize very restricted portions of the antigen molecule. The part of the antigen interacting with the B or T cell receptor is called an *antigenic determinant*, or *epitope* (Figure 3.1). Thus, antigens are molecules bearing one or more epitopes which, by definition, may be recognized by specific receptors in an immune system. The appropriate descriptive adjective is *antigenic*, while the property described is *antigenicity*.

Immunogenicity (similarly, *immunogen, immunogenic*) is the property of stimulating the immune system to generate a specific immune response. Strictly speaking, immunogens are a subset of antigens. Thus, all immunogens are antigens, but not vice-versa. Although the immune system may *recognize* an antigen, it does not *respond* to it unless it is also immunogenic. However, the terms antigen and immunogen are very often used interchangeably.

Some antigens may render the immune system unable to generate a response to that particular antigen, while the ability to respond to other antigens is preserved. The condition of antigen-specific immune unresponsiveness is called *tolerance* (*tolerogen, tolerogenic, tolerogenicity*). The properties of immunogenicity and tolerogenicity are not necessarily mutually exclusive for a given substance. Depending on the antigen dose and route of administration, its physical and chemical characteristics,

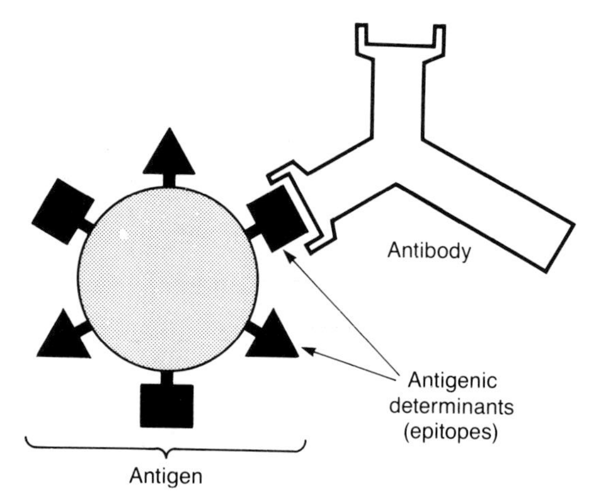

Figure 3.1. *Antigenic determinants interact with immune system receptors.* Although antigens and B and T cell receptors are often of similar size, the area of interaction between the two entities is limited to a small portion of the surfaces of the two molecules. The parts of an antigen molecule making contact with immune system receptors is called an *antigenic determinant* or *epitope.* In the diagram, an antigen molecule is shown interacting with an antibody molecule (B cell receptor). This particular antigen is *multivalent,* it is trivalent with respect to each of two different epitopes (triangles and squares). The interaction of T cell receptors with antigens is quite different from that depicted here (see text and Chapter 7 for details).

and the presence of other substances influencing immune system activation or function, the same material may be immunogenic or tolerogenic. Tolerance will be discussed in detail in Chapter 9.

Antigenic determinants

A single antigen molecule often displays more than one antigenic determinant, or epitope. These may be structurally identical or not. The number of epitopes present on a single molecule is called the *valence.* Antigens with one epitope are *monovalent,* while antigens with several determinants are *multivalent* or *multideterminant* (Figure 3.1).

Epitopes recognized by antibodies

Epitopes may be formed by primary, or by higher order macromolecular structure. A *sequential* epitope is determined by primary structure. In a protein, this type of epitope is determined by a segment of the amino acid sequence (Figures 3.2 and 3.3). Small polypeptides corresponding to portions of larger proteins have been used extensively to analyze protein-antibody interactions.

The B subunit of cholera toxin is a pentamer of polypeptides each 103 amino acids long. A peptide corresponding to amino acids 50–64 of this protein reacts well with antisera obtained after immunization of rabbits with whole cholera toxin. Furthermore, antibodies generated by immunization with the peptide inhibit the biological activity of cholera toxin, that is, these antibodies *neutralize* the toxin. Thus, the portion of the B subunit protein from amino acids 50 to 64 is a sequential epitope of cholera toxin.

A. Sequential epitope

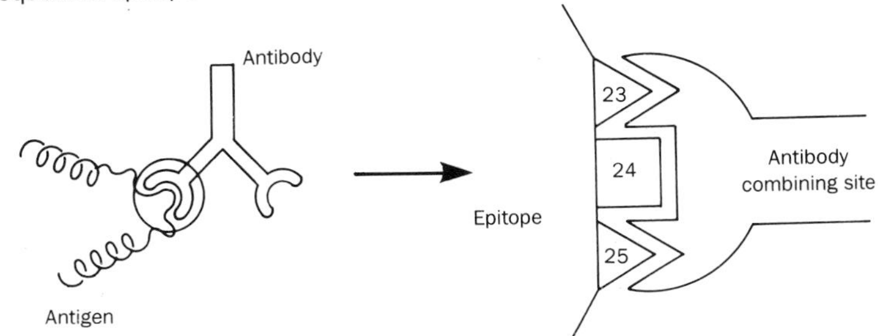

B. Conformational epitope

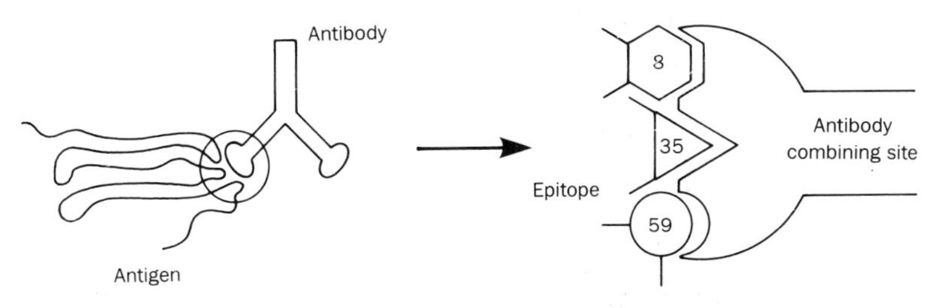

Figure 3.2. *Sequential and conformational epitopes.*

A. *Sequential epitope.* Here an antibody interacts with a portion of antigen that is loosely folded and has little secondary structure. At the right of the figure the interaction is diagrammed in more detail. Three consecutive (hence, sequential) residues contribute to the epitope bound by the antibody.

B. *Conformational epitope.* Here the antibody interacts with a region of antigen involved in higher order macromolecular structure. Three residues which are far apart in the primary sequence are brought close together to form an epitope. Clearly, the integrity of this epitope depends on secondary and tertiary structure.

Some epitopes may be hidden in the three dimensional structure of a native protein and become exposed only after physical or chemical alterations in conformation. On the other hand, epitopes may also be formed by portions of the molecule distant in the primary structure, but juxtaposed in secondary or tertiary structure. In this case, the epitope's integrity may be lost when the macromolecule is denatured. Since these epitopes depend on the conformation of the macromolecule, they are often referred to as *conformational antigenic determinants* or *conformational epitopes* (Figure 3.2).

An X-ray crystallographic analysis of the complex of an antibody fragment with the antigen hen egg lysozyme (HEL) showed that 16 amino acids of HEL made contact with the antibody (Table 3.I). These residues are located in two portions of the primary sequence widely separated from one another. This is an example of a conformational epitope.

Some epitopes may be more immunogenic than other determinants present on the same molecule. Hence, a preponderance of antibodies specific for these *immunodominant* epitopes of an antigen may be produced in a humoral immune response. Consider the dengue virus which causes epidemics of a hemorrhagic fever syndrome in tropical regions. The envelope glycoprotein of this virus is 495 amino acids long. Hence, 490 peptides containing six amino acids can be generated from its sequence. When these peptides were reacted with sera from patients after infection with dengue (or closely related) viruses, 22 peptides distributed in seven distinct domains reacted with all sera. Thus, these peptides define the immunodominant epitopes in the humoral response to dengue virus envelope glycoprotein (Figure 3.3).

With respect to the humoral response, immunodominant epitopes are usually exposed on the antigen surface, and are frequently rich in

Table 3.I. RESIDUES COMPRISING A CONFORMATIONAL EPITOPE OF HEN EGG LYSOZYME

#	Residue	#	Residue
18	Asp	116	Lys
19	Asn	117	Gly
21	Arg	118	Thr
22	Gly	119	Asp
23	Tyr	120	Val
24	Ser	121	Gln
25	Leu	124	Ile
27	Asn	129	Leu

Data from Bentley et al., 1989.

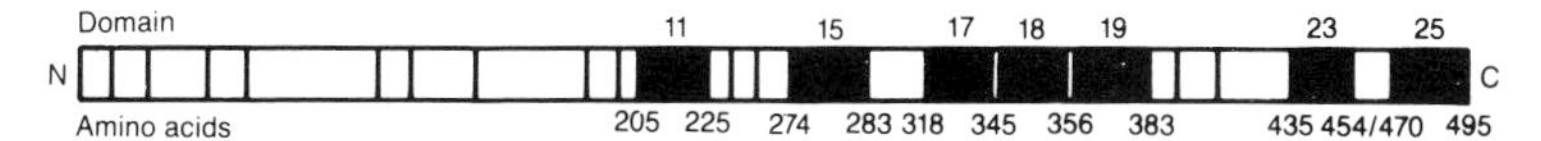

Figure 3.3. *Immunodominant epitopes of the dengue virus envelope glycoprotein.* The 490 hexapeptides spanning the sequence of a dengue virus envelope glycoprotein were reacted with sera from infected humans. Twenty two peptides defining seven domains (of a total 25) reacted with all sera. Thus, these domains (black in the diagram) may be considered immunodominant in the humoral response to dengue virus. The amino (N) and carboxyl (C) termini of the protein are indicated. (Data from Innis et al., 1989.)

electric charges (hydrophilic). Amino acids with charged side chains at physiologic pH are lysine, arginine, histidine, glutamate and aspartate.

Epitopes recognized by T cells

As will be seen in later chapters, there are many similarities in the structure of antibodies and T cell antigen receptors. There is, however, an extremely important difference in the ways in which these two types of receptor bind antigen. B cell receptors (antibodies) bind to intact antigen molecules in solution or adsorbed onto cells or other surfaces. T cell receptors, on the other hand, react only with antigen *fragments* (peptides) associated with special cell-surface glycoproteins (*histocompatibility proteins*) on cell surfaces (see Chapter 7).

Since T cells recognize antigen fragments, T cell epitopes are of the sequential type. Immunodominant T cell epitopes appear to be regions of a protein which are amphipathic (have both hydrophilic and hydrophobic residues) and which are capable of forming helical secondary structure. Examples of this are two regions of the *Plasmodium falciparum* circumsporozoite protein. The two immunodominant T cell epitopes of this protein are amino acids 326-345 and 361-380.

Haptens

Haptens are molecules reacting with specific antibodies (they are antigenic) but they are not immunogenic unless they are coupled to a *carrier*. Landsteiner observed that low-molecular weight polysaccharides interacted with antibodies produced in response to immunization with polysaccharide-protein complexes, while injection of polysaccharide alone produced no response. Since the polysaccharide component of the complex determined some of the antibody specificities, Landsteiner called the sugars *haptens* (from the Greek *hapto* meaning "to grasp"). This work led to many investigations of artificial immunogens produced by conjugating small (non-immunogenic) molecules of known structure with carrier proteins. Three types of antibody specificity may be obtained subsequent to immunization with hapten-carrier conjugates: specificity for the hapten, for the carrier, and for new antigenic determinants created by the union of hapten and carrier (Figure 3.4). Antibodies specific for the hapten will react with it even when it is attached to a different carrier. In addition, the binding of hapten-specific antibodies to the complex can be inhibited by free hapten.

Many naturally-occurring molecules may be viewed as hapten-carrier conjugates. The hapten phosphorylcholine (PC) is found on the surfaces of many microbes. Examples include the bacteria *S. pneumoniae*, and the filaria *W. bancrofti*. Some antibodies produced in individuals infected with these agents bind to PC and will react with it when it is coupled to a carrier entirely unrelated to microbial components.

Non-immunogenic chemicals may react with protein components of the skin, creating new immunogenic structures. The immune system may respond to these new antigens resulting in a contact hypersensitivity reaction (see Chapter 7). An example of this is the catechol derivative urushiol found in certain plants of the genus *Toxicodendron*, more affectionately known as poison ivy.

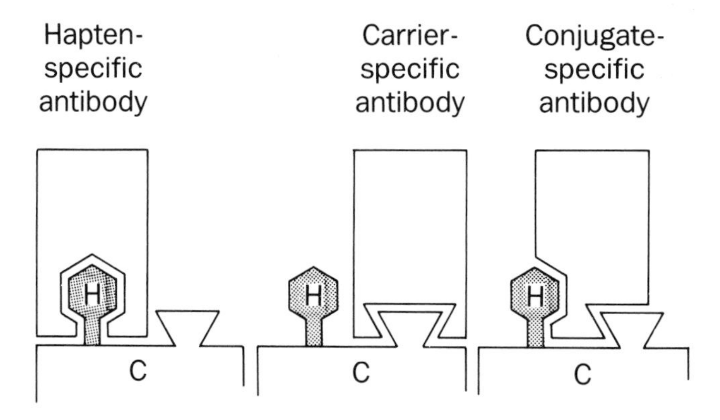

Figure 3.4. *Epitopes of a hapten-carrier conjugate.* Three types of epitopes may be distinguished by antibodies reacting with a hapten-carrier conjugate: epitopes determined solely by the hapten; epitopes found only on the carrier; and new epitopes created by the coupling of hapten and carrier with contributions from both molecules.

Specificity and cross-reactivity

The specificity of a B cell or T cell receptor for a particular antigenic determinant is relative, not absolute. That is, the receptor is capable of binding several antigenic determinants, albeit all very similar in molecular structure (Figure 3.5). A single epitope may also react with several receptors differing only slightly in structure. Thus, we may say that

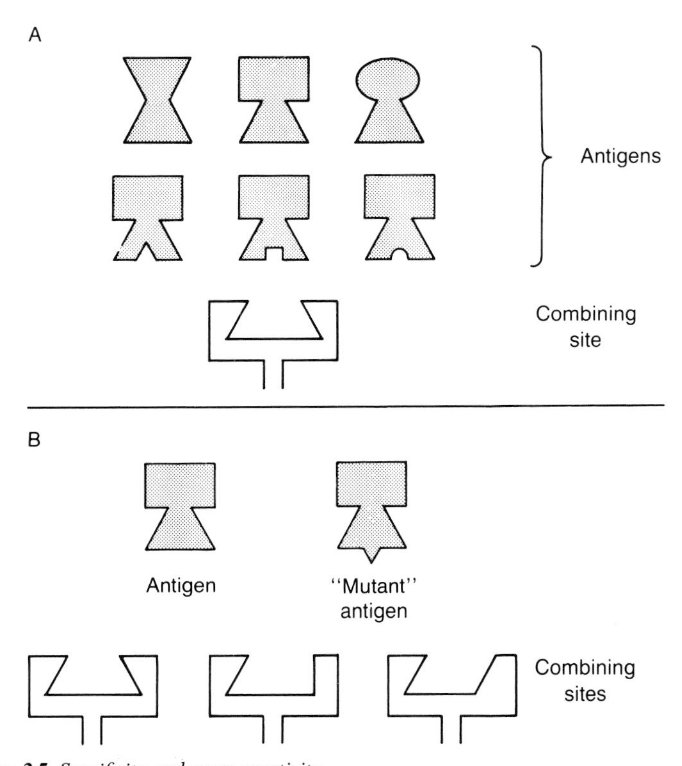

Figure 3.5. *Specificity and cross-reactivity.*

A. *Cross-reactive antigens.* Several different antigens may react with the same B cell receptor (antibody). The first three antigens in the figure are all different from one another, but bind to the same receptor because they all have an epitope in common. The next set of antigens all have epitopes which are slightly different from "a perfect fit," yet still capable of binding to the receptor.

B. *Cross-reactive receptors.* These three receptors, despite differences in their structures, still react with the same epitope. However, a small change in the epitope ("mutation") may abolish its interaction with the receptor. (Note: the concepts of specificity and cross-reactivity apply to the interaction of epitopes with T cell receptors also. However, this interaction is different from that depicted here, see Chapter 7 for details).

a group of molecules bearing subtly different epitopes are *cross-reactive* with respect to a particular receptor, or that a group of subtly different receptors are cross-reactive with respect to a given epitope. These subtle differences in the structures of epitopes and receptors are reflected in the different affinities or "strengths" of their interactions (see Chapter 5).

While certain small changes in the structure of an epitope may still permit interaction with the same receptor (with different affinities), other changes may entirely abolish the interaction (Figure 3.5). For example, some antibodies binding influenza hemagglutinin (HA) are *neutralizing* antibodies since they may block attachment of the virus to its receptors and prevent infection. Culturing influenza viruses in the presence of HA-specific antibodies allows one to select virus variants having mutant HA. These variants have point mutations in the HA gene resulting in one or a few amino acid substitutions. Mutations destroying or creating glycosylation signal sequences may also affect HA structure and antigenicity. These small changes may allow the virus to escape the neutralizing effect of antibodies in culture.

Factors influencing immunogenicity

Humoral versus cell-mediated immunity

Antibodies and T cell antigen receptors have different requirements for their interactions with antigen. Humoral responses are directed against intact antigens. Cell-mediated responses require association of antigen fragments with cell membrane proteins. The process whereby soluble antigens become membrane-associated, and how this affects humoral and cellular immunity is discussed in Chapters 6 and 7.

Foreignness

An important factor determining immunogenicity is the degree of difference between the antigen and similar entities within the host. The most immunogenic substances are those which have no homologues in the responding organism. Foreignness is often thought of as the "evolutionary distance" between two species. Thus, a mouse will generate an immune response to rat hemoglobin and *vice-versa*, but both will respond more vigorously to immunization with bacterial proteins. Since mice and rats are both mammals having speciated relatively recently in evolution, many of their proteins are similar. The proteins of the more primitive bacteria differ much more from those in either rodent than do rat and mouse proteins from each other. An animal normally does not develop an immune response against self components which may be immunogenic in other species.

Size

The molecular weight of a substance is also important in determining immunogenecity. As a general rule, substances with higher molecular weight are more immunogenic. For example, somatotropin, $M_r = 45{,}000$ daltons, is more immunogenic than adrenocorticotrophic hormone, $M_r = 4{,}600$. Proteins with $M_r < 5$–6,000 are usually not immunogenic, however, exceptions occur. For example, glucagon with $M_r = 2{,}600$ is a good immunogen. Large molecular size alone is insufficient to confer immunogenicity, however. Large synthetic polymers such as polyacrylamide, nylon, and teflon are not immunogenic. In general, polysaccharides are not very immunogenic until they reach $M_r = 20$–30,000. Dextran (a polymer of glucose), with M_r ranging from 40,000 to 40,000,000 is a potent immunogen.

Higher order molecular structure

Immunogenicity is enhanced by complexity due to secondary, tertiary or quaternary molecular structure. As mentioned above, immunodominant T cell epitopes appear to have mainly helical secondary structure. Physical or chemical denaturation, the breaking of disulfide bonds, or disruption of non-covalent interactions altering conformation or unfolding polypeptides decreases immunogenicity. Small changes in structure may not change immunogenicity, but may alter antigenicity, as with influenza virus hemagglutinin, discussed above. Some antibodies binding metmyoglobin (containing heme) do not bind apomyoglobin (without heme). The heme group influences overall myoglobin structure.

Large polysaccharides may also have secondary structure contributing to immunogenicity.

Dosage

An antigen may be immunogenic at one dose, while being non-immunogenic or even tolerogenic at another. Some substances, like the SIII polysaccharide of streptococci, induce tolerance at low concentrations (*low-zone tolerance*). Other materials may induce tolerance at high concentrations (*high-zone tolerance*, see Chapter 9).

Age

Polysaccharides from *S. pneumoniae*, *N. meningitidis*, or *H. influenzae* fail to induce an immune response in infants. However, the same materials evoke strong responses in adults. This phenomenon results from the immaturity of the neonate's immune system which has not yet developed sufficient numbers of all of the specialized cell types required for efficient response to many antigen challenges.

Genetic composition of the host

Differences in the immunogenicity of a particular antigen may also occur in association with well-defined differences in particular genetic loci between individuals. Thus, *responder* and *low-* or *non-responder* phenotypes may be distinguished. These phenotypes are determined by *immune response genes* (*Ir genes*) which are part of the *major histocompatibility complex* (*MHC*, see Chapter 8). Ir genes influence the magnitude and quality of the immune response. For example, human T cell responses to schistosomal antigens appear to be controlled by genes within the DR and DQ loci of the MHC.

Functional classification of antigens

Substances inducing humoral immune responses do not all stimulate the immune system in precisely the same manner. Some antigens interact predominantly with B cells, while others require various cooperative mechanisms between B cells, T cells, and other cell types. Antigens able to stimulate B cells directly are called *T-independent* (*TI*) antigens, while those that require T and B cell cooperation are called *T-dependent* (*TD*) antigens. TI antigens are further subdivided into *TI type 1* (or simply, *type 1*) and *TI type 2* (*type 2*) antigens.

T-independent antigens are large molecules made up of repeating identical units. Thus, they display many identical epitopes. Type 1 antigens are also called *polyclonal B cell activators* or *B cell mitogens* (agents inducing mitosis). Examples of this type of antigen are *lipopolysaccharides* (*LPS*) from the cell wall of gram negative bacteria, or *Nocardia water-soluble mitogen* from the cell wall of the actinomycetes genus *Nocardia*. These substances can stimulate B cells to secrete antibodies *irrespective of the specificity of those antibodies.*

Many polysaccharides of bacterial or plant origin are type 2 antigens. Examples are levans (fructose polymers), galactans (galactose polymers), and dextrans (glucose polymers). Polymerized proteins such as flagellin (a building block of bacterial flagellae) may also be type 2 antigens.

Early research with athymic or "nude" mice (so-called because they lack fur) showed that they did not produce antibodies against most protein and glycoprotein antigens. Nude mice have no thymus and, thus, no mature T cells, but have fully functional B cells. Antigens not eliciting antibody responses in nude mice were called thymus- (T-) dependent for this reason. The vast majority of proteins and glycoproteins encountered in nature are TD antigens. Humans with thymic defects cannot produce antibodies against TD antigens. Examples of this condition include the DiGeorge and Nezeloff syndromes. The differences in the antibody responses to TI and TD antigens are discussed in Chapter 6.

Adjuvants

Adjuvants are substances conferring immunogenicity upon, or increasing the immunogenicity of other antigens. Adjuvants may affect macrophages, causing them to secrete factors stimulatory for lymphocytes, or they may directly stimulate lymphocytes themselves. There is no single chemical characteristic common to all adjuvants. Precipitating proteins with inorganic substances such as alum or beryllium salts may increase their immunogenicity. Alum precipitates or diphtheria and tetanus toxoids are used for vaccination.

The most powerful adjuvants are of bacterial origin, particularly cell wall components of the family *Actinomycetes* including the *Mycobacteria*, the *Nocardiae*, and the *Corynebacteria*. Dead mycobacterial cells admixed with mineral oil constitute *Freund's complete adjuvant* (*FCA*). This is a very potent adjuvant capable of augmenting both humoral and cell-mediated immune responses. Unfortunately, these natural adjuvants have only limited use in humans since they have severe side effects such as lethargy or fever, or may activate autoimmune processes (immune responses against self components, see Chapter 10). FCA alone may cause granuloma formation or a condition known as *adjuvant arthritis*. Many investigators are searching for natural or synthetic materials which will be potent adjuvants lacking harmful side effects.

As discussed in Chapter 2, our immune systems have cellular elements mediating both specific and non-specific protective mechanisms. We have described in this chapter some of the important characteristics of macromolecules in our environment in the context of specific immunity. However, we also possess protective mechanisms lacking the ability to make the extremely subtle discriminations of structure inherent in specific immune responses. These mechanisms may act independently, or in concert with specific immune responses, and are the subject of the following chapter.

SOURCES AND SUGGESTED ADDITIONAL READING

Barret, D. J. (1985) Human immune responses to polysaccharide antigens: an analysis of bacterial polysaccharide vaccines in infants. *Adv. Pediatr.*, **32**:139–158.

Bentley, G. A., Alzari, P. M., Amit, A. G., Boulot, G., Guilon-Chitarra, V., Fischmann, T., Lascombe, M.-B., Mariuzza, R. A., Poljak, R. J., Riottot, M.-M., Saul, F. A., Souchon, H. & Tello, D. (1989) Studies of structure and specificity of some antigen-antibody complexes. *Phil. Trans. R. Soc. Lond. B*, **323**:487–494.

Berzovsky, J. A., Cease, K. B., Cornette, J. L., Spouge, J. L., Margalit, H., Berkower, I. J., Good, M. F., Miller, L. H. & DeLisi, C. (1987) Protein antigenic structures recognized by T cells: potential applications to vaccine design. *Immunol. Rev.*, **98**:9–52.

Crumpton, M. J. (1966) Conformational changes in sperm-whale myoglobin due to combination with antibodies to apomyoglobin. *Biochem. J.*, **100**:223–232.

DeGroot, A. S., Johnson, A. H., Maloy, W. L., Quakyi, I. E., Riley, E. M., Menon, A., Banks, S. M., Berzovsky, J. A. & Good, M. F. (1989) Human T cell recognition of polymorphic epitopes from malaria circumsporozoite protein. *J. Immunol.*, **142**:4000–4005.

Hirayama, K., Matsushita, S., Kikuchi, I., Iuchi, M., Ohta, N. & Sasazuki, T. (1987) HLA-DQ is epistatic to HLA-DR in controlling the immune response to schistosomal antigen in humans. *Nature*, **327**:426–430.

Innis, B. L., Thirawuth, V. & Hemachudha, C. (1989) Identification of continuous epitopes of the envelope glycoprotein of dengue type 2 virus. *Am. J. Trop. Med. Hyg.*, **40**: 676–687.

Jacob, C. O., Sela, M. & Arnon, R. (1983) Antibodies against synthetic peptides of the B subunit of cholera toxin: crossreaction and neutralization of the toxin. *Proc. Natl. Acad. Sci. USA*, **80**: 7611–7615.

Lal, R. B. & Ottesen, E. A. (1989) Antibodies to phosphocholine-bearing antigens in lymphatic filariasis and changes following treatment with diethylcarbamazine. *Clin. Exp. Immunol.*, **75**: 52–57.

Landsteiner, K. (1936) *The Specificity of Serological Reactions.* Charles C. Thomas, Springfield, Illinois. (Harvard University Press, Cambridge, revised edition, 1945.)

Musher, D. M., Chapman, A. J., Goree, A., Jonsson, S., Briles, D. & Baughn, R. E. (1986) Natural and vaccine-related immunity to *Streptococcus pneumoniae. J. Infect. Dis.*, **154**: 245–256.

Warren, H. S., Vogel, F. R. & Chedid, L. A. (1986) Current status of immunological adjuvants. *Ann. Rev. Immunol.*, **4**: 369–388.

Webster, R. G., Laver, W. G. & Air, G. M. (1983) Antigenic variation among type A influenza viruses. In: *Genetics of Influenza Viruses*, P. Palese & D. W. Kingsbury eds., Springer-Verlag, New York, pp. 127–168.

Chapter 4

Non-Specific Immunity

In our discussion, we will divide non-specific immune processes into three broad categories: tissues and secretions preventing entry of microorganisms into the body; humoral factors killing or inhibiting the growth of pathogens; and cells ingesting and destroying microbes.

PREVENTING ENTRY

The simplest way to avoid infection is to prevent microorganisms from gaining access to the body. Major lines of defense are the skin and epithelia lining inner body surfaces such as the respiratory and gastrointestinal mucosae (Figure 4.1). The skin is both a mechanical and biochemical barrier. Keratinized squamous epithelial cells of the epidermis form a tough flexible sheet covering the body. In addition, perspiration and sebaceous secretions contain lactic and fatty acids producing a low pH environment at the skin surface, inhibiting microbial growth. Furthermore, continual desquamation of surface epithelial cells eliminates microbes which have adhered to the skin.

The protection of mucosal surfaces differs from that of the skin. Epithelium lining the respiratory tract does not have a strong keratinized covering, and is constantly moistened with fluid (mucus) at physiologic pH. However, mucus itself has antimicrobial properties. Mucus is rich in *sialoproteins*, glycoproteins containing sialic acid. Many cell-surface glycoproteins also contain sialic acid, and some of these serve as receptors for viral attachment, a notable example being the influenza virus. Sialoproteins in mucus compete for viral binding sites and confer some protection against infection.

In the trachea and bronchi, secreted mucus is swept upward by epithelial cilia at a rate of 1–2 cm/min. Microbes and other particles trapped in the mucus are swept into the pharynx and exporated or swallowed. Cigarette smoke and some viral infections (e.g., influenza)

A

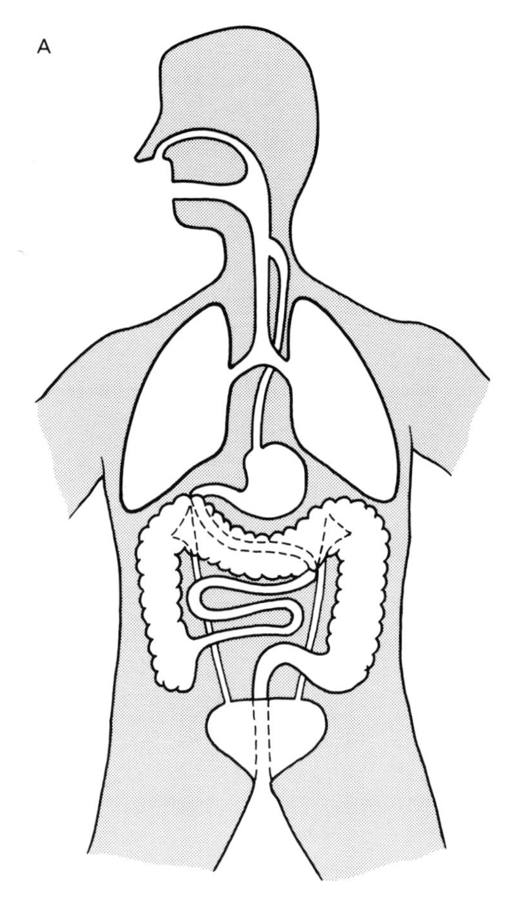

B

SKIN	RESPIRATORY OR GI TRACT
almost anything	air or enteric fluid
perspiration, sebum	mucus
keratinized epidermis	bronchial epithelium, alveoli
	gastrointestinal epithelium
non-keratinized epidermis	basement membrane
interstitial fluid	
capillary endothelium	
LYMPH/BLOOD	

cause ciliary paralysis and stagnation of mucus. This leads to an increased incidence of respiratory tract infections. In *Kartagener's syndrome*, (chronic sinusitis, bronchiectasis and situs inversus) there is an increased incidence of respiratory infections. This disease has autosomal dominant inheritance; it results from a microtubule defect and is a form of ciliary dyskinesis. An abnormal form of the microtubule-binding ATPase *dynein* has been implicated in the etiology of this condition.

Lysozyme is an enzyme found in many body secretions (tears, perspiration, mucus, saliva), as well as in blood, fluids bathing internal body cavities, and in the lysosomes of phagocytic cells (neutrophils, macrophages). Thus, lysozyme is a component of all three categories of non-specific immunity. This enzyme cleaves $\beta(1-4)$ glycosidic bonds linking N-acetylglucosamine and muramic acid in bacterial cell walls. The bacteria are thus rendered more fragile with respect to osmotic and oncotic perturbations in their environment, and more susceptible to bactericidal substances produced by phagocytic cells.

Although the mildly alkaline pH of saliva inhibits growth of oral flora, conditions significantly antagonistic to bacterial multiplication

---

◀

Figure 4.1. *The human body's interfaces with the environment.*

A. *Body topology.* We are accustomed to thinking of the boundary between the inside and outside of our bodies as being the skin. However, the "interiors" of several organs are actually topologically external to the body's tissues. A small indigestible object when swallowed may pass unimpeded through the gastrointestinal tract and out through the anus without penetrating a membrane. Similarly, foreign bodies may pass in and out of the respiratory and genitourinary systems without crossing barriers. The epithelia lining these spaces (as well as the duct systems of the exocrine glands secreting their products into them), in addition to the skin, form the boundaries between the environment and the tissues and fluids regulated by the body's homeostatic mechanisms.

B. *Characteristics of the interfaces.* Once a microbe gains access to the blood, it has the potential to travel to any area of the body. What are the physical barriers which must be crossed in order for this to happen? This depends on the route being traveled by the invader, i.e., the skin vs. a mucosal surface.

The skin comes into contact with a multitude of substances: dirt, water, soap, lotions, creams, etc. In addition, sweat and sebaceous glands secrete their own products (which have antimicrobial properties, see text) onto its surface. Keratinized epithelium is the outermost layer of the skin, non-keratinized cells immediately underneath. Interstitial fluid and capillary endothelium remain to be traversed before the blood is reached.

Entering from a mucosal surface such as the lung or GI tract, the first barrier encountered is mucus which contains secreted antibodies and other antimicrobial substances (see text). The epithelium and its basement membrane must then be penetrated before encountering the interstial fluid and capillary endothelium and, finally, the blood.

are not encountered in the digestive system until the stomach is reached. Gastric pH is frequently as low as 1.0; this is lethal to most microorganisms. As are upper respiratory airways, the digestive tract is coated with a mucus film inhibiting microbial penetration of underlying tissues.

Acidophilic organisms, or those protected from gastric acid (inside food, for example) survive transit through the stomach and enter the intestines, an ideal environment for bacterial growth. Fortunately, the lower GI tract is heavily colonized with non-pathogenic *normal flora* inhibiting colonization and growth of potential pathogens in a variety of ways. Normal flora compete for nutrients and attachment sites in the intestines, and produce toxic metabolic waste products. Some organisms even produce antibiotics which may inhibit multiplication of pathogens.

As described in Chapter 2, much lymphoid tissue underlies the epithelia of the respiratory and GI tracts. These tissues produce antibodies which are secreted into the lumen by epithelial cells (see Chapter 5). Thus, the mucus contains both nonspecific (e.g., sialoproteins and lysozyme) and specific (antibodies) antimicrobial substances.

Epithelia of the urinary tract are also coated by mucus, and periodic flushing with moderately acidic urine helps to keep this system bacteria-free. In the male reproductive system, semen contains bacteriostatic polycations such as spermine and spermidine, as well as a bactericidal concentration of zinc. In women, cells underlying the epithelium of the vagina store large quantities of glycogen. As these cells die and desquamate, the normal vaginal flora metabolize the glycogen to lactic acid, creating a bacteriostatic environment.

HUMORAL FACTORS

The complement system

The complement system is a group of serum proteins with diverse biological effects which can be divided into four categories: destruction of microbes and cells (*lysis*); attraction of leukocytes to areas of infection and inflammation (*chemotaxis*); facilitation of the phagocytosis of microbes (*opsonization*); and stimulation of inflammation and hypersensitivity (*anaphylatoxicity*). Complement components circulate in the blood in an inactive form. Under certain conditions, a complex cascade of enzymatic reactions leads to the successive activation of each component in the pathway, very much as in blood coagulation. Table 4.I summarizes the major characteristics of complement components, fragments, regulatory proteins, and receptors. The word complement is often abbreviated as *C′*, which is understood to include *all* components of the complement system.

Most complement components are synthesized by hepatocytes. Recently, it has also been shown that mononuclear phagocytes are able to synthesize all complement proteins. Additional sources, such as intestinal epithelia, of some components have also been identified. Probably there are yet other sources of complement awaiting discovery.

There are two interrelated pathways of complement activation, the *classical pathway*, and the *alternate pathway*. Some components are used in only one pathway, others participate in both.

The classical pathway of complement activation

This cascade of reactions is diagrammed in Figure 4.2.

For the most part, the nomenclature of complement components in the classical pathway reflects the order of their activation in the cascade. Each component is designated by the letter *C* followed by a number, for example, *C3*. The *C1 complex* consists of three distinct proteins: *C1q*, *C1r*, and *C1s*. Their activated forms are designated by asterisks (e.g., *C1r**). With the remainder of the numbered components (*C2–C9*), addition of a lowercase letter (e.g., *C3a*, *C3b*) designates an active fragment generated by proteolysis of the inactive precursor form (C3). Many active fragments are highly unstable and degrade spontaneously or through the action of inactivators (proteases) present in serum. These inactive fragments are designated with the letter i before the name of the fragment (e.g., *iC3b*).

Antigen-antibody complexes initiate the classical pathway. Consider the situation where antibody is bound to a cell-surface antigen. The C1 complex comprised of one molecule each of C1q and C1s, and two molecules of C1r, is called the *recognition unit*. C1q mediates binding of the recognition unit to antibody-antigen complexes on the cell surface. In order to initiate the cascade, C1q must bind simultaneously to two antibodies. Two different types of antibody, IgG and IgM (see Chapter 5) may bind complement. IgG antibodies are monomers, thus, two IgG molecules must bind to the cell surface in very close proximity in order to permit interaction with C1q. On the other hand, most

Table 4.I. CHARACTERISTICS OF COMPLEMENT COMPONENTS AND FRAGMENTS

Name	M_r (kd)	Serum conc. (mg/dl)	Function
Classical Pathway			
C1 complex	820		
C1q	400	10	Binds to antibody-antigen complexes
C1r	170		Activates C1s
C1s	80	12	Cleaves C2 and C4
C2	100		
C2a			Forms C3 convertase with C4b
C2b			
C3	185	120	
C3a			Anaphylatoxin, opsonin
C3b			Forms C5 convertase with C4b2a
C3c			
C3d			B cell growth factor
C4	240	25	
C4a			
C4b			Forms C3 convertase with C2a
C5	170	10	
C5a			Anaphylatoxin
C5b			Nucleus for lytic unit
C6	100	60	Component of lytic unit
C7	110	50	Component of lytic unit
C8	160	50	Component of lytic unit
C9	70	30	Component of lytic unit
Alternate pathway			
Factor B	93	20	
Ba			
Bb			Forms C3 convertase with C3b
Factor D	25	0.1	Cleaves B to Ba and Bb
Properdin	185	2	Stabilizes C3bBb
Regulatory proteins			
Soluble:			
C1INH	90		Blocks activity of Cls*
C3bINA	90		Cleaves C3b to C3c and C3d
Factor H	150		Necessary for activity of C3bINA
C4BP	250		Allows C3bINA to cleave C4b
Cell surface:			
DAF	70		Promotes dissociation of C3 and C5 convertases
MCP			Binds C3b, iC3b, enhances activity of factor I
HRF	65		Binds C8, C9, prevents formation of lytic unit
Receptors			
CR1			Opsonic C3b and iC3b receptor
CR2			C3d and Epstein-Barr virus receptor
CR3			Binds iC3b
CR4			Binds iC3b

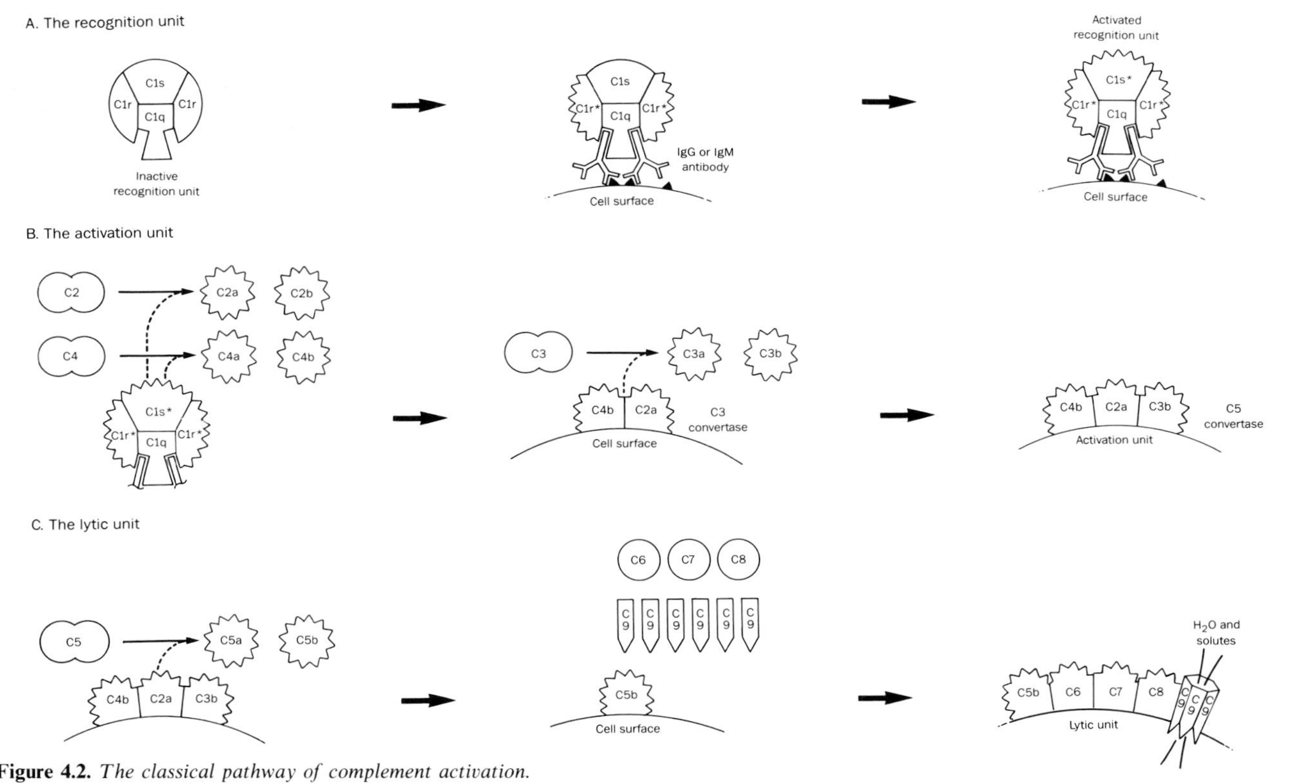

Figure 4.2. *The classical pathway of complement activation.*

A. *The recognition unit.* The inactive C1 complex, or recognition unit, binds to the C regions of two antibodies adherent to a cell. This binding activates the recognition unit.

B. *The activation unit.* This sequence of reactions generates the C3 and C5 convertase complexes. The latter is the activation unit.

C. *The lytic unit.* C5b forms the nucleus of the lytic unit which is completed by the successive binding of C6, C7, C8 and several molecules of C9, forming a membrane channel.

circulating IgM is pentameric (five antibody molecules joined together). Consequently, a single molecule of pentameric IgM bound to a surface may initiate the classical pathway. C1q binding allosterically activates C1r to cleave itself into its active form (C1r*) which modifies C1s to its active form (C1s*).

The next group of components in the sequence forms the *activation unit*. C1s* cleaves C4 yielding C4a and C4b, and C2 yielding C2a and C2b. Since one recognition unit may cleave many molecules of C2 and C4, this is the first amplification step in the cascade. C4b and C2a join to form the complex C4b2a, often referred to as *C3 convertase*, which then associates with the cell membrane independently of the recognition unit. Cell-bound C4b2a then cleaves C3 yielding C3a and C3b. C4b2a also amplifies the reaction cascade because it may cleave many molecules of C3. C3b then associates with membrane-bound C4b2a to form the complete activation unit, C4b2a3b. This complex is also called *C5 convertase* since it cleaves C5 yielding C5a and C5b. This is also an amplification step since many molecules of C5 may be cleaved by one molecule of C5 convertase.

The final steps in the classical pathway form the *membrane attack unit*, or *lytic unit*. C5b associates with the cell membrane, independently of the recognition and activation units, and is a nucleus for the lytic unit. In succession, one molecule each of C6, C7, and C8 combine with the membrane-bound C5b. Several molecules of C9 complete the lytic unit. Formation of this complex requires only non-covalent association of components without enzymatic modification. The completed membrane attack unit is a lesion in the cell membrane allowing free flow of water and solutes leading to osmotic lysis. The C9 component of the lytic unit closely resembles proteins mediating T cell cytotoxicity (see Chapter 8).

Some of the C3b formed in this series of reactions may associate with the cell membrane independently of C4b2a. Cell-bound C3b is an *opsonin*, that is, it enhances the ingestion of the cell by phagocytic cells (neutrophils, macrophages, etc., see below) bearing receptors for C3b. When these receptors bind C3b, the phagocytic process is activated (see Figure 4.5).

The C3a and C5a complement fragments are potent *anaphylatoxins*. These substances stimulate release of histamine from mast cells causing a localized increase in vascular permeability and contraction of smooth muscle. Additional effects are release of hydrolytic enzymes from neutrophils, and aggregation of platelets. This results in microthrombosis, stasis of blood flow and accumulation of extravascular fluid (edema), as well as localized tissue destruction. In addition to being an anaphylatoxin, C5a is a *chemotactic factor* for phagocytic cells. When PMNs and macrophages come into contact with C5a, their random motion decreases. This causes them to accumulate at areas with high concentrations of C5a.

The alternate pathway of complement activation

This series of reactions is summarized in Figure 4.3.

The components unique to the alternate pathway of complement are called *factor B*, *factor D*, and *properdin.*

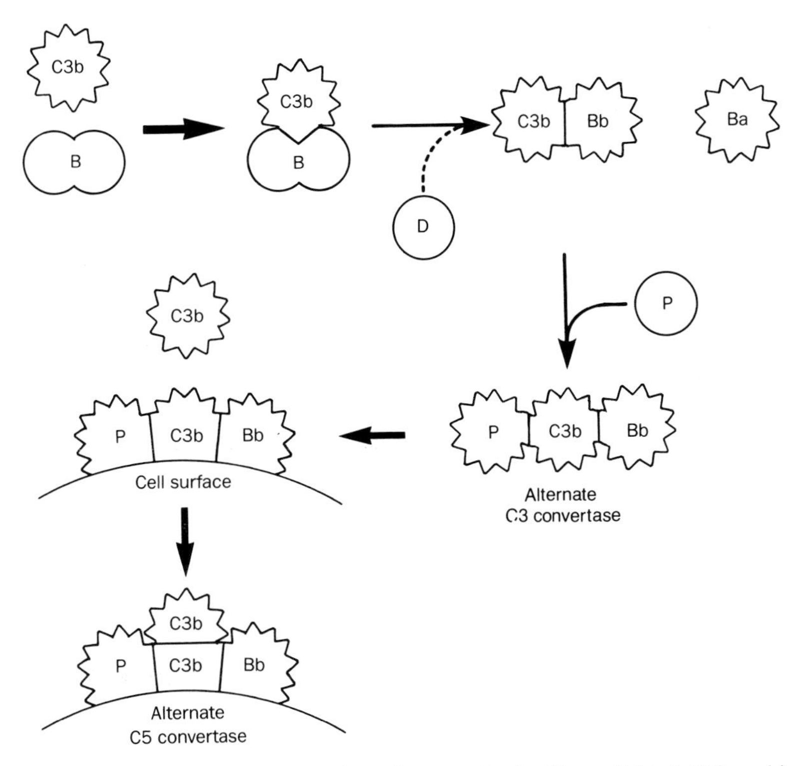

Figure 4.3. *The alternate pathway of complement activation.* Factor B binds C3b, and is then a substrate for factor D which generates the unstable alternate pathway C3 convertase C3bBb. When stabilized by properdin (P), the complex may bind to cells. Addition of one more molecule of C3b to the complex forms the alternate pathway C5 convertase. Subsequent reactions occur as diagrammed in Figure 4.1 C.

Normal serum contains very low concentrations of C3b. This probably arises via the action of serum proteases on circulating C3. *Factor B* binds C3b yielding the C3bB complex. This is the substrate for *factor D*. The factor B in the C3bB complex is cleaved yielding Ba and Bb, the complex now being C3bBb. This is a proteolytic complex which cleaves C3 to C3a and C3b analogously to the classical pathway complex C4b2a (it is the C3 convertase of the alternate pathway). This complex is highly unstable and loses activity rapidly unless it is further complexed with *properdin* to give PC3bBb.

Although the complex PC3bBb is continually generated in very small amounts, the alternate pathway is said to be initiated when this complex is protected from dissociation and inactivation by adhering to certain surfaces. The polysaccharides in the glycolipids and glycoproteins on microbial cells are very effective stabilizers of PC3bBb. When the complex binds to a microbe's surface, it catalyzes production of larger and larger amounts of C3b. When another molecule of C3b associates with PC3bBb to give $P(C3b)_2Bb$, the complex becomes the C5 convertase of the alternate pathway. Subsequent to generation of C5b, the lytic unit forms as described above for the classical pathway.

It is evident from the above description of the classical and alternate pathways of complement activation that they are interconnected through components C3 and C5, and the remainder of the lytic unit (Figure 4.4). Although these mechanisms may operate independently of one another, they probably are activated together in most instances. Remember, however, that the classical pathway requires antibodies for its initiation. In their absence, only the alternate pathway is active. Note that because these mechanisms share C3 and C5, the important anaphylatoxins C3a and C5a are generated even if only one pathway is operating.

Regulatory complement components

Several complement components act to suppress the activation of this system. Four of these regulators circulate in the blood, as do the components described above (Table 4.I, Figure 4.4). *C1 esterase inhibitor* (*C1INH*) blocks the proteolysis of C2 and C4 by C1s* (a component of the recognition unit) in the classical pathway. *C3b inactivator* (*C3bINA* or *factor I*), as its name implies, inactivates C3b in conjunction with the cofactor *C3 binding protein* (*factor H*, also called *β1H*). The binding of factor H to C3b within complexes promotes their dissociation

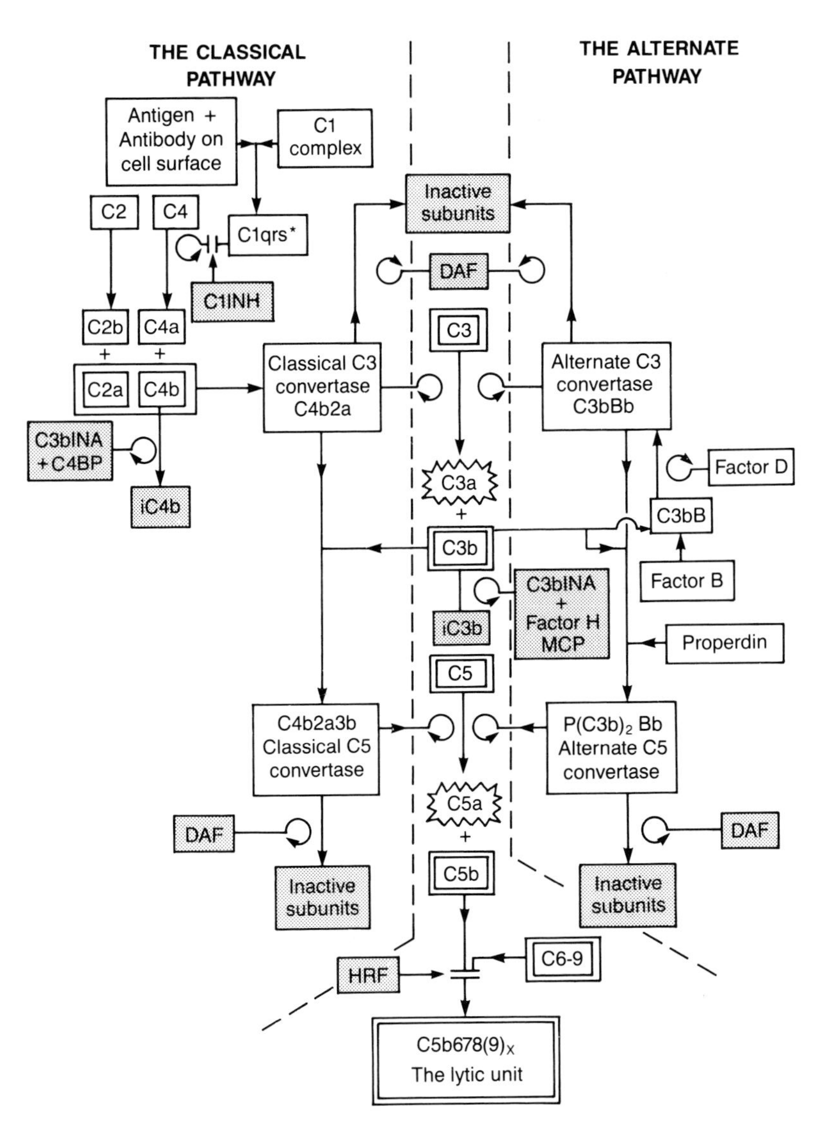

Figure 4.4. *Interconnections of the classical and alternate pathways of complement activation.* This figure diagrams the classical and alternate pathways of complement activation side-by-side, and shows the components they have in common. These are C3, C5 and the remainder of the lytic unit (as well as some regulatory proteins). Circular arrows indicate catalytic (enzymatic) processes carried out by the indicated components or complexes. Gray boxes indicate regulatory components or inactive fragments. Note that regardless of which pathway of complement is activated, the important anaphylatoxins C3a and C5a are generated.

into inactive subunits. The action of C3INA yields *iC3b*, which may be active as an opsonin (see below), and further breakdown products, some of which have additional biologic activities within the immune system. For example, the fragment *C3d* is a growth factor for B cells. *C4 binding protein* (*C4BP*) binds to C4b and renders it susceptible to inactivation by C3INA.

Three additional cell membrane components act to protect cells from complement lysis. *Decay accelerating factor* (*DAF*) is a protein anchored in the cell membranes of erythrocytes, leukocytes, platelets, and endothelial cells. This protein interacts with complement complexes deposited on the cell's surface and, as its name suggests, hastens their dissociation (inactivation). *Membrane cofactor protein* (*MCP*) is found on leukocytes and platelets. It binds to C3b and iC3b and appears to enhance the activity of factor I. *Homologous restriction factor* (*HRF*) has a distribution similar to that of DAF. This protein binds to C8 and C9 and inhibits formation of an effective lytic unit.

Complement receptors

Four receptors for complement components and fragments have been identified and partially characterized with respect to their distribution, structure, and possible roles in immune system function (Table 4.I).

CR1 (also called CD35, see Chapter 2 for an explanation of the CD nomenclature) on erythrocytes, B cells, macrophages and mesangial cells, is a receptor for C3b and iC3b. CR1 mediates the opsonic actions of C3b and iC3b. *CR2* (a.k.a. CD21) of B cells is the receptor for C3d which stimulates B cell proliferation. CR2 is also the receptor for the Epstein-Barr virus which causes infectious mononucleosis and has been associated with Burkitt's lymphoma (see Chapter 10). Two other complement receptors, *CR3* and *CR4* have recently been identified on phagocytic cells. Both of these receptors bind iC3b. CR3 is comprised of two chains. The α chain is also called CD11b, the β chain is CD18. CR3 has been shown to bind to a number of microbial cell surface constituents as well as to several leukocyte cell surface proteins known as "adhesion molecules" (see Chapter 7). Thus, this receptor may have one or more roles in addition to its binding of iC3b. The biological consequences of CR3 and CR4 receptor occupancy are not yet well-characterized.

We have seen that the complement system is an important non-specific humoral defense mechanism. It effectively endows antibodies with the potential to kill the cells to which they bind (classical pathway); it binds directly to and lyses microorganisms (alternate pathway); it enhances the phagocytosis of microbes; and it generates a series of secondary products with a variety of pharmacological properties resulting in enhanced antibody responses, increased numbers of leukocytes at sites of infection, and inflammatory phenomena.

Genetic defects of the complement system

Genetically-determined deficiency of one or more complement proteins occurs in association with two principal types of clinical syndromes. One is autoimmune disease (see Chapter 10), and the other is recurrent bacterial and/or fungal infections. Most complement component genes are codominantly expressed. Thus, heterozygotes for a defective gene have serum component levels on the order of 50% of normal (clinically silent), while homozygotes have 0–10%. Table 4.II lists syndromes associated with complement deficiency.

Not all individuals with deficiency of a particular component will have recurrent microbial infections. The additional factors determining the severity of expression of a component deficiency are not known. Similarly, the relationship between particular component deficiencies and the origins of immune responses against self is unknown.

Table 4.II. SYNDROMES OF COMPLEMENT COMPONENT DEFICIENCY

Deficient component(s)	Associated Clinical syndromes
C1	SLE, recurrent bacterial infections
C2	SLE, HSP, DMI, recurrent spesis, meningitis, pneumonia, esp, pneumococci
C3	SLE, recurrent bacterial infections
C4	SLE, sepsis and meningitis (rare)
C5	SLE, sepsis, esp. *Neisseria*
C6, C7, or C8	Sepsis, esp. *Neisseria*
C9	Usually no disease, occasionally as for C5-8
Factor D	Recurrent sinusitis, bronchitis
Properdin	Meningococcal meningitis
C1INH	Hereditary angioedema
CR1	SLE

Abbreviations: DMI, dermatomyositis; HSP, Henoch-Schönlein purpura; SLE, systemic lupus erythematosus. (Adapted from Johnston, 1989.)

C3 deficiency frequently results in recurrent bacterial pneumonia, meningitis or peritonitis. The organisms most often responsible for these infections are *S. pneumoniae*, *S. aureus*, and *S. meningitidis*. Deficiency of any of the components C5–C8 also results in recurrent bacterial infections, *Neisseria* species account for most of these cases. C9 deficiency is most often not associated with any pathology, although some individuals have increased susceptibility to *Neisseria*.

Hereditary angioedema results from decreased production of C1 esterase inhibitor, or production of a defective protein. The absence of C1INH leads to unrestrained proteolysis of C2 and C4 by activated C1s*. The precise mechanism leading to tissue edema is unknown. A vasoactive fragment of C2 may be responsible. The disease is characterized by periodic acute subepithelial swelling, involving predominantly the skin of the extremities, and the gastrointestinal mucosa. Swelling in the oropharynx or brain may be fatal due to asphyxiation or cerebral edema. Inheritance is autosomal dominant.

Acute phase proteins

Certain serum proteins are called acute phase proteins because their concentrations rise dramatically during microbial infections and other inflammatory processes (the *acute phase reaction*). Most acute phase proteins are synthesized by hepatocytes. *C-reactive protein* (*CRP*) is a serum of β-globulin which binds to microorganisms whose cell walls contain phosphocholine (e.g., the C substance of streptococci). CRP binding to these surfaces may stabilize PC3bBb and initiate the alternate pathway of complement activation. Other acute phase proteins are summarized in Table 4.III. While the concentrations of the acute phase proteins in inflammation and infection increase, the concentrations of

Table 4.III. ACUTE PHASE PROTEINS

Protein	Function
C-reactive protein (CRP)	Enhances chemotaxis, phagocytosis
α_1 proteinase inhibitor (α_1 antitrypsin)	Proteinase inhibitor
α_1 acid glycoprotein (AGP, orosomucoid)	Inhibits platelet activation, expressed on lymphocytes
Fibrinogen	Coagulation
C3	Complement component
Factor B	Complement component

other serum proteins may decrease. Examples are albumin and transferrin. The physiology of the acute phase reaction and its role in controlling inflammation or infection is not well-understood.

PHAGOCYTIC CELLS

Phagocytosis of microbes and pinocytosis of their products (e.g., toxins) are indispensable components of non-specific defense. Phagocytosis is ingestion of particulate matter, while pinocytosis is uptake of extracellular fluid and macromolecules in solution. Neutrophils, eosinophils, monocytes and macrophages are cells carrying out these processes.

In addition to eliminating invading microorganisms, phagocytic cells (especially macrophages) remove old and damaged serum proteins and cells from the circulation. This is an important "housekeeping" function, clearing our bodies of cellular debris and defunct proteins.

The complex process of phagocytosis may be divided into several stages: *chemotaxis*, *attachment*, *ingestion*, *degranulation*, *intracellular killing*, and *intracellular digestion* (Figure 4.5).

Chemotaxis

The encounter of phagocytic cells with their microbial prey is not entirely random. Various molecules of leukocyte or bacterial origin may orient phagocyte movement toward areas of infection and inflammation. This process of chemical direction of leukocyte movement is called *chemotaxis*. Substances inducing phagocyte accumulation are called *chemotactic factors*.

Upon death, bacteria release several substances chemotactic for neutrophils and macrophages. Two of these are small peptides, formyl-Met-Leu-Phe (FMLP) and formyl-Met-Phe. Some complement fragments and complexes, e.g., C5a and C5b67, also have chemotactic activity. Chemotactic complement fragments may be generated independently of activation of the entire complement cascade. Bacterial proteases, as well as some present in serum (trypsin, plasmin) and others released from leukocytes, act directly on C5 to form C5a.

Leukocytes themselves release a number of factors chemotactic for other leukocytes. Mast cells and basophils release *eosinophil chemotactic*

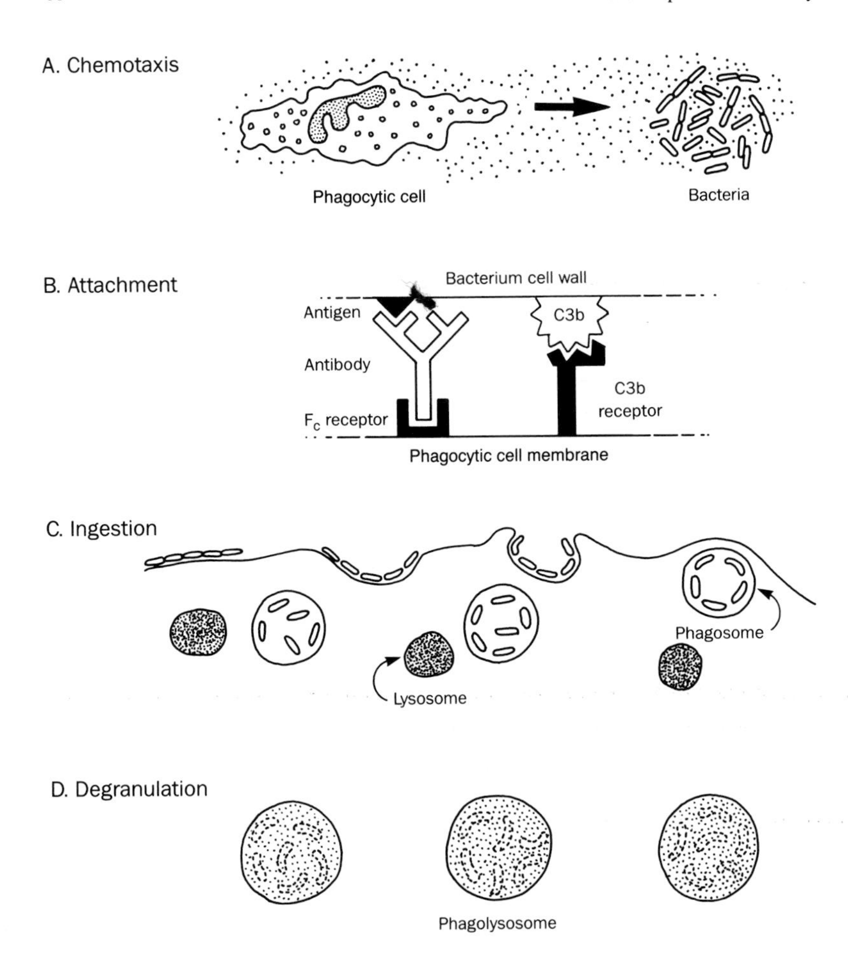

Figure 4.5. *Phagocytosis.*

A. *Chemotaxis.* Bacterial products and leukocyte factors modulate phagocytic cell motility. Eventually, these cells aggregate at areas of infection and/or inflammation, where the concentration of chemotactic factors is high.

B. *Attachment.* Phagocytic cells possess membrane receptors for complement components and antibody. Occupancy of these receptors by molecules that are also bound to a cell (e.g., a bacterium) surface enhances phagocytosis. This phenomenon is called *opsonization.*

C. *Ingestion.* The phagocyte membrane reorganizes and engulfs particles adherent to its surface. This process generates membrane-bound intracellular vesicles (*phagosomes*) containing the particles.

D. Degranulation. Lysosomes fuse with phagosomes, creating phagolysosomes, and the macromolecular constituents of the ingested particles are degraded.

factor of anaphylaxis (*ECF-A*), a mediator of the immediate hypersensitivity reaction (see Chapter 10). Mast cell granules also contain *neutrophil chemotactic factor*. Enzymes such as kallikrein, fibrin degradation products, and some prostaglandins may also be chemotactic for neutrophils and monocytes.

Chemotactic factors bind to specific receptors on phagocyte surfaces and induce an energy-dependent reorganization of cytoskeletal components. The end result is an ameoboid progression of the leukocyte in a direction of increasing concentration of the chemotactic factor.

Chemotaxis may be observed experimentally in a *Boyden chamber*. This device consists of two compartments separated by a microporous membrane. Leukocytes are placed into one compartment while a chemotactic substance is added to the other. The experiment is scored by measuring the fraction of leukocytes migrating through the membrane pores during a given period of time (a few hours).

Attachment

Attachment is the adherence of particles to phagocyte cell membranes. *Opsonization* is the process by which certain proteins facilitate attachment. The complement fragment C3b and the IgG class of antibodies (see Chapter 5) have this property. Phagocytic cells have receptors for IgG antibodies and C3b (Figure 4.6). IgG or C3b bound to a microbe acts as a bridge between the bacterial and phagocytic cells (Figure 4.5). Phagocyte C3b or IgG receptor occupancy triggers the next phase of phagocytosis.

Ingestion

During engulfment, pseudopodia actively reach out to enclose particles. The movement of the cell membrane requires reorganization of cytoskeletal components (microtubules and microfilaments) and is energy-dependent (uses ATP). The cell membrane completely surrounds the particles and fuses forming an intact outer membrane and a phagocytic vesicle (*phagosome*) in the cytoplasm (Figures 4.5).

Following ingestion, a number of changes in cellular metabolism occur. O_2 consumption and lactic acid production increase. Glucose

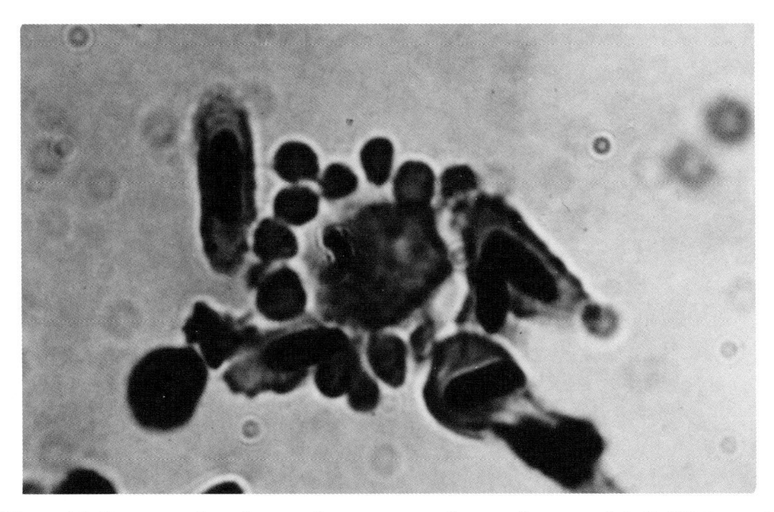

Figure 4.6. *Demonstration of macrophage receptors for complement and IgG.* Chicken red blood cells (nucleated) were incubated with rabbit IgG anti-chicken erythrocyte antibodies. Ox red blood cells (anucleate) were incubated with rabbit IgM anti-ox erythrocyte antibodies and complement. Macrophages do not bear Fc receptors for IgM. The murine macrophage in the center binds to complement fixed by IgM on ox RBCs, and to IgG on chicken RBCs. Clearly, the rabbit proteins are sufficiently homologous to their murine counterparts to interact with these macrophage receptors.

metabolism increases, especially the hexose monophosphate shunt, rising from 1% to 10% of total glucose metabolism. Membrane lipid synthesis also increases.

Degranulation

Degranulation is the discharge of cytoplasmic granule (lysosome) contents into phagosomes without subjecting the phagocyte's cytoplasm to the injurious effects of degradative enzymes contained within. Phagosomes and lysosomes approximate one another in the cytoplasm, then fuse their membranes creating a *phagolysosome* (Figure 4.5).

Intracellular killing

Within 30–60 minutes after engulfment, most microorganisms are inactivated. If the phagocytic cells are disrupted with detergents or an osmotic gradient, and the previously ingested microorganisms placed in a nutrient medium, they fail to grow. Microbes are killed in phagolysosomes by toxic metabolites, and lysosome contents. Antimicrobial mechanisms may be classified as *oxygen-dependent* or *oxygen-independent*.

Oxygen-dependent microbicidal mechanisms

A consequence of the increased O_2 consumption and oxidative metabolism (*the respiratory burst*) accompanying engulfment is increased output of potently microbicidal oxygen metabolites: *hydrogen peroxide* (H_2O_2), *superoxide anion* (O_2^-), *hydroxyl radicals* ($OH\cdot$), and *singlet oxygen.* Although phagocytosis may occur during anaerobiosis, efficient intracellular killing partly depends on increased oxidative metabolism.

The *myeloperoxidase-*H_2O_2*-halide* system generates potent antimicrobial hypohalous acids (e.g., HClO) from H_2O_2 and halide (predominantly chloride and iodide) ions. The H_2O_2 in these reactions may be bacterial in origin, for example, streptococci themselves produce H_2O_2. Other bacteria, e.g., staphylococci, synthesize the enzyme *catalase* which generates O_2 and H_2O from H_2O_2. Hydrogen peroxide cannot accumulate when catalase-producing organisms are ingested.

Oxygen-independent microbicidal mechanisms

These lysosome constituents are summarized in Table 4.IV.

The *defensins* are the most recently discovered nonspecific antimicrobial peptides isolated from phagocytes. Three such peptides have

Table 4.IV. ANTIMICROBIAL LYSOSOME CONSTITUENTS

Compound	Functions
Acid hydrolases (proteases, nucleases)	Hydrolytic enzymes with low pH optima, degrade ingested macromolecules
Bactericidal/permeability-increasing factors (cationic proteins)	Increase permeability of bacterial cell walls
B_{12}-binding protein	Inhibits B_{12}-dependent enzymes
Defensins	?
Lysozyme	Muramidase, degrades bacterial cell walls
Lactoferrin	Chelates Fe, inhibits Fe-dependent enzymes
Neutral proteases	Degrade proteins

been found in humans. They are 29–30 residues long, and rich in cysteine, arginine, and aromatic amino acids. These peptides are very similar to each other, as well as to defensins isolated from granulocytes of other species. Defensins have anti-bacterial, anti-fungal, and virus-neutralizing activity *in vitro* against some organisms. The extent of their function in these capacities *in vivo* is not yet known.

Phagocyte deficiencies

Phagocyte deficiencies commonly affect more than one leukocyte lineage. However, primary deficiencies of one cell lineage can exist with normal or even increased numbers of others.

The number of granulocytes in the blood of healthy individuals fluctuates between 2–9,000/mm^3. Levels may increase during infection or injury (such as a burn), or may decrease in other situations (e.g., drug-induced bone marrow suppression). A decreased number of circulating leukocytes is called *leukopenia*, while a decrease in granulocytes is *granulocytopenia*. Lack of neutrophils is *neutropenia*, and absence of granulocytes from blood *and* bone marrow is *agranulocytosis*. Individuals with neutropenia often have increased susceptibility to infections with pygenic bacteria. Several genetic defects may result in granulocytopenia.

Congenital chronic neutropenia is the name given to a heterogenous group of genetically-determined disorders characterized by a low rate of production of mature neutrophils. The most common complications are skin or respiratory tract infections with *S. aureus*, or overgrowth or dissemination of normal enteric flora. The frequency and severity of these infections correlate with the degree of neutropenia.

In *cyclic neutropenia*, as its name implies, the number of circulating neutrophils fluctuates. Oral or skin infections may occur during episodes of neutropenia. A stem cell defect may be responsible.

Acquired neutropenia may have several causes. In some instances, pregnant women produce antibodies to surface components of the fetus's granulocytes. These antibodies cause premature neutrophil demise (*isoimmune neonatal neutropenia*). Some drugs complex with granulocyte membrane glycoproteins and induce antibody formation against these complexes. These antibodies may also decrease granulocyte life span. Other drugs may cause neutropenia by interfering with hemopoiesis. Cancer chemotherapeutic agents, for example, frequently cause bone marrow suppression. Some individuals have idiosyncratic reactions to certain drugs (e.g., chloramphenicol) resulting in neutropenia secondary to myelosuppression. These situations are referred to as *drug-induced neutropenia.*

Phagocyte defects

Some genetic defects do not affect the number of cells produced, but impair their function. These defects may be categorized by the particular aspects of phagocytosis affected.

In the *lazy leukocyte syndrome*, granulocytes exhibit impaired chemotaxis and motility. This is due to an abnormal distribution of actin filaments within phagocytic cells. These patients have increased susceptibility to bacterial infections, especially agents invading the skin and mucous membranes.

Several diseases result from an inability of phagocytes to kill ingested microbes. *Chronic granulomatous disease (CGD)* is a heterogeneous group of disorders which have been classified into four types based on their patterns of inheritance and molecular defects (Table 4.V). A defective cytochrome b has been found in the majority of patients. This abnormality impairs the generation of antimicrobial oxidizing substances (H_2O_2, etc.) during phagocytosis.

CGD may manifest itself in infancy with disseminated abscesses and lymphadenopathy (enlarged lymph nodes) and splenomegaly (enlarged spleen), or chronic dermatitis. These patients later manifest characteristic suppurative granulomas localized in the lungs, spleen, bones, and other areas.

Glucose-6-phosphate dehydrogenase deficiency also leads to impaired intracellular killing, since glucose oxidation via the hexose monophosphate shunt is reduced, and H_2O_2 production decreased. Patients with this defect exhibit increased susceptibility to bacterial infections.

Table 4.V. CHRONIC GRANULOMATOUS DISEASE

Type	Mode of inheritance	Molecular defect
Type I	X-linked recessive	Large subunit of cytochrome b not synthesized
Type Ia	X-linked recessive	Large subunit of cytochrome b is defective
Type II	Autosomal recessive	?
Type III	Autosomal recessive	Defective cytochrome b (different from Type Ia)
Type IV	X-linked recessive	?

Chediak-Higashi disease is an autosomal recessive syndrome of a phagocyte defect associated with partial albinism, photophobia, nystagmus, mental retardation, peripheral neuropathy, and a defect in natural killer cell activity. Neutrophils characteristically show abnormally large ("giant") lysosomes. Phagocytes are able to ingest and kill only small numbers of bacteria. The primary abnormality appears to be in degranulation, but chemotactic function is also altered. These patients have increased susceptibility to pyogenic infections and lymphoma.

Myeloperoxidase deficiency impairs function of the MPO-H_2O_2-halide system. These patients have increased susceptibility to fungal and bacterial infections.

Neutrophil specific granule deficiency may be either acquired or inherited (autosomal recessive). As its name states, this syndrome is characterized by absence of neutrophil specific granules. These granules contain (among other things) lactoferrin and B_{12}-binding protein. Affected individuals are susceptible to recurrent bacterial infections.

A syndrome consisting of chronic skin abscesses, sinopulmonary infections, coarse facies and elevated serum IgE is known as *hyper-immunoglobulin E syndrome.* Despite very high levels of serum IgE, these patients do not exhibit allergic phenomena (see Chapter 10). This appears to be a heterogeneous group of disorders. Some patients have defects in phagocyte chemotaxis, others have no (as yet) identifiable abnormalities of phagocyte function.

Older children and adolescents may be afflicted with *localized juvenile periodontitis.* In this condition, recurrent severe oral infections lead to marked loss of alveolar bone and premature loss of teeth. Neutrophils in these patients lack a surface glycoprotein that has not yet been characterized.

Malakoplakia is a rare acquired macrophage defect. Inflammatory granulomas are prominent in various tissues (epithelia of the urinary tract, most commonly). These lesions contain large mononuclear cells with mineralized aggregates of bacteria in phagosomes (Michaelis-Gutman bodies). A defect in the degradation of ingested bacteria is postulated.

The frequently severe pathology resulting from the malfunction of the humoral and cellular mechanisms of non-specific immunity argue strongly for their importance in effective resistance to microbial infection.

Now we shift our focus to the mechanisms of specific immunity. We are only beginning to understand the tremendous complexity and subtlety of the interplay of cells and soluble factors in these mechanisms. We begin by examining the molecules through which specific immunity was discovered: antibodies.

SOURCES AND SUGGESTED ADDITIONAL READING

Bignold, L. P. (1988) Measurement of chemotaxis of polymorphonuclear leukocytes *in vitro*. The problems of the control of gradients of chemotactic factors, of the control of the cells, and of the separation of chemotaxis from chemokinesis. *J. Immunol. Methods*, **108**: 1–18.

Boesen, P. (1988) Cyclic neutropenia terminating in permanent agranulocytosis. *Acta Med. Scand.*, **223**: 89–91.

Curnutte, J. T., ed. (1988) Phagocytic Defects I: Abnormalities Outside of the Respiratory Burst. *Hematol. Oncol. Clin. North Am.*, **2** (1).

Curnutte, J. T., ed. (1988) Phagocytic Defects II: Abnormalities of the Respiratory Burst. *Hematol Oncol. Clin. North Am.*, **2** (2).

Coonrod, J. D. (1986) The role of extracellular bactericidal factors in pulmonary host defense. *Semin. Respir. Infect.*, **1**: 118–129.

Curran, F. T. (1987) Malakoplakia of the bladder. *Br. J. Urol.*, **59**: 559–563.

Dierich, M. P., Schulz, T. F., Eigentler, A., Huemer, H. & Schwäble, W. (1988) Structural and functional relationships among receptors and regulators of the complement system. *Mol. Immunol.*, **25**: 1043–1051.

Dowton, S. B. & Colten, H. R. (1988) Acute phase reactants in inflammation and infection. *Semin. Hematol.*, **25**: 84–90.

Fair, W. R. & Parrish, R. F. (1981) Antibacterial substances in prostatic fluid. *Prog. Clin. Biol. Res.*, **75A**: 247–264.

Fearon, D. T. (1988) Complement, C receptors, and immune complex disease. *Hosp. Pract. (Off.)*, **23**: 63–72.

Fries, L. F., Siwik, S. A., Malbran, A. & Frank, M. M. (1987) Phagocytosis of target particles bearing C3b-IgG covalent complexes by human monocytes and polymorphonuclear leucocytes. *Immunology*, **62**: 45–51.

Goldman, J. M., Foroozanfar, N., Gazzard, B. G. and & Hobbs, J. R. (1984) Lazy leukocyte syndrome. *J. R. Soc. Med.*, **77**: 140–141.

Johnson, E. & Hetland, G. (1988) Mononuclear phagocytes have the potential to synthesize the complete functional complement system. *Scand. J. Immunol.*, **27**: 489–493.

Johnston, R. B., Jr. (1989) Disorders of the complement system. In E. R. Stiehm, ed., *Immunologic Disorders in Infants and Children*, Third Edition, W. B. Saunders Company, Philadelphia, pp. 384–399.

Levine, D. H. & Madyastha, P. R. (1986) Isoimmune neonatal neutropenia. *Am. J. Perinatol.*, **3**: 231–233.

Melchers, F., Erdei, A., Schulz, T. & Dierich, M. P. (1985) Growth control of activated, synchronized murine B cells by the C3d fragment of human complement. *Nature*, **317**: 264–267.

Pincus, S. H., Boxer, L. A. & Stossel, T. P. (1976) Chronic neutropenia in childhood. Analysis of 16 cases and a review of the literature. *Am. J. Med.*, **61**: 849–861.

Rolfe, R. D. (1984) Interactions among microorganisms of the indigenous intestinal flora and their influence on the host. *Rev. Infect. Dis.*, **6**:S73–79.

Snyder, J. D. & Walker, W. A. (1987) Structure and function of intestinal mucin: developmental aspects. *Int. Arch. Allergy Appl. Immunol.*, **82**:351–356.

Wanner, A. (1986) Mucociliary clearance in the trachea. *Clin. Chest Med.*, **7**:247–258.

Young, G. A. & Vincent, P. C. (1980) Drug-induced agranulocytosis. *Clin. Haematol.*, **9**:483–504.

Chapter 5

Immunoglobulins

Electrophoresis of serum proteins yields four major peaks (Figure 5.1). Albumin is the most abundant, normally comprising 50–70% of total serum protein. The other three are the *alpha*, the *beta*, and the *gamma globulins*. The gamma globulins are now called *immunoglobulins* (often abbreviated *Ig*), or *antibodies* (*Ab*). These are large, complex glycoproteins sharing a common structural motif, and grouped into classes and subclasses according to variations upon this motif.

For the moment, all one need know about antibody protein structure is that a molecule of Ig contains two different polypeptides called *heavy chains*, and *light chains*.

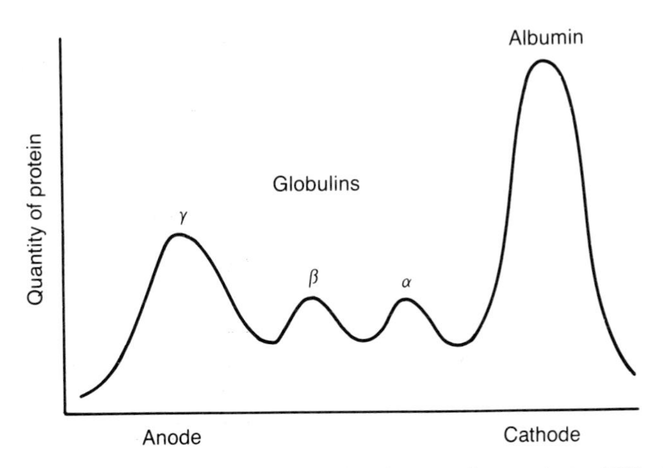

Figure 5.1. *Serum protein electrophoresis.* Albumin normally constitutes 65% of total serum protein. The remaining proteins are classified as α, β, or γ globulins based on electrophoretic mobility. Antibodies (immunoglobulins) are γ globulins.

IMMUNOGLOBULIN GENES

Organization

There are several different classes of Ig categorized by heavy chain structure (to be described below). There are also two different types of antibody light chains, κ (*kappa*) and λ (*lambda*). The genes encoding antibody heavy and light chains are located on different chromosomes. Table 5.I shows the locations of these gene complexes.

A single gene encoding a complete antibody heavy or light chain does not exist as such within the DNA of most cells. These genes are assembled by the union of separate gene segments. These segments are widely separated in germ cells and all somatic cells, except B lymphocytes in which they *rearrange* their relative positions to create a "mature" antibody gene encoding a functional protein. This remarkable pattern of somatic gene rearrangement has been demonstrated in only one other system, the genes encoding the antigen receptors of T cells. As will be seen shortly, this rearrangement process is the core of the immune system's ability to recognize the tremendous variety of antigenic structures in nature.

An Ig light chain gene is assembled from three types of gene segments. These are the *light chain variable region* (V_L), *joining region* (J_L), and *constant region* (C_L) gene segments. Similarly, heavy chain gene segments are V_H, J_H and C_H, and another type of gene segment called D (for *diversity*). The nomenclature of the gene segments derives from studies of the amino acid sequences of antibody heavy or light chains. Thus, the variable region has its name because this part of the molecule shows great variation from one antibody chain to another. The constant region, as its name implies, varies little. The joining and diversity regions were the names given to areas interposed between V and C, each having its own characteristic sequence pattern. It is the variable regions of the heavy and light chains together which constitute the antibody *combining site* (also called the *paratope*). This is the portion of the Ig molecule which makes contact with antigen (an epitope).

Table 5.I. CHROMOSOMAL LOCATIONS OF IMMUNOGLOBULIN GENES

Locus	Chromosome
Heavy chain	14
κ light chain	2
λ light chain	22

The overall organization of heavy and light chain genes is very similar. The κ locus on chromosome 2 consists of from 100–300 V_κ genes, followed by five J_κ genes, and one C_κ gene (Figure 5.2). Similarly, the heavy chain locus on chromosome 14 contains about 100 V_H genes, approximately 30 D genes, six J_H genes and 11 C_H region genes: μ (*mu*), δ (*delta*), $\gamma 3$ (*gamma*-3), $\gamma 1$, ψ_ε (*pseudo-epsilon*), $\alpha 1$ (*alpha*-1), ψ_γ (*pseudo-gamma*), $\gamma 4$, $\gamma 2$, ε (*epsilon*), and $\alpha 2$ (Figures 5.3 and 6.5). Each of these C_H genes corresponds to a particular immunoglobulin class (see below). The human λ locus on chromosome 22 is somewhat different, being composed of several (an unknown number) V_λ, six J_λ and six C_λ genes (Figure 5.4).

V_H and V_L genes are grouped into *V gene families* based on nucleotide or amino acid sequence homology. That is, V genes > 80% homologous are said to belong to the same family. Homology between families is generally < 75%. Human V_H genes have been divided into six families designated V1–V6. Similarly, human V_κ genes have been divided into 4 families, V_λ genes into six.

A typical V gene consists of two exons separated by an intron. Exons are DNA sequences that are translated into protein. Intron sequences, intervening between exons, are transcribed into messenger RNA, but

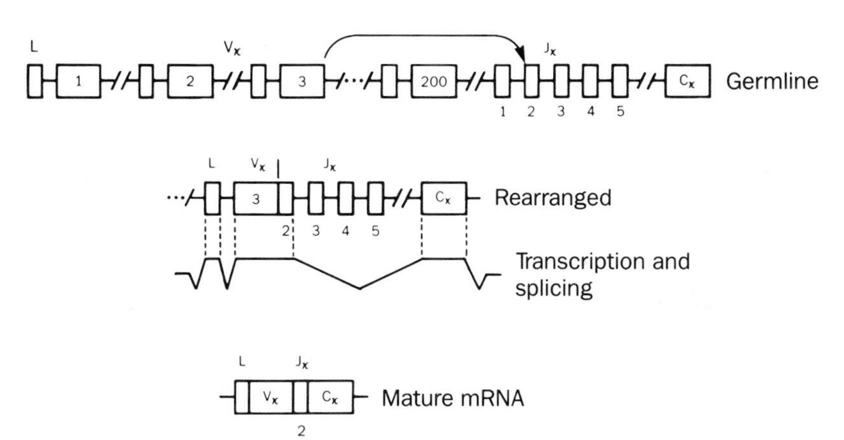

Figure 5.2. *Organization, rearrangement, and transcription of human immunoglobulin κ chain genes.* By convention, diagrams such as this are drawn with left to right corresponding to the direction of mRNA transcription, 5′ to 3′. This also corresponds to the direction of protein synthesis, beginning with the amino terminus and ending with the carboxyl terminus. In the germline configuration, a few hundred V_κ genes are arranged 5′ to the five J_κ genes and the single C_κ gene. A single DNA rearrantement event joints one V_κ gene to one J_κ gene. After transcription, introns are spliced out yielding mature mRNA encoding a κ light chain.

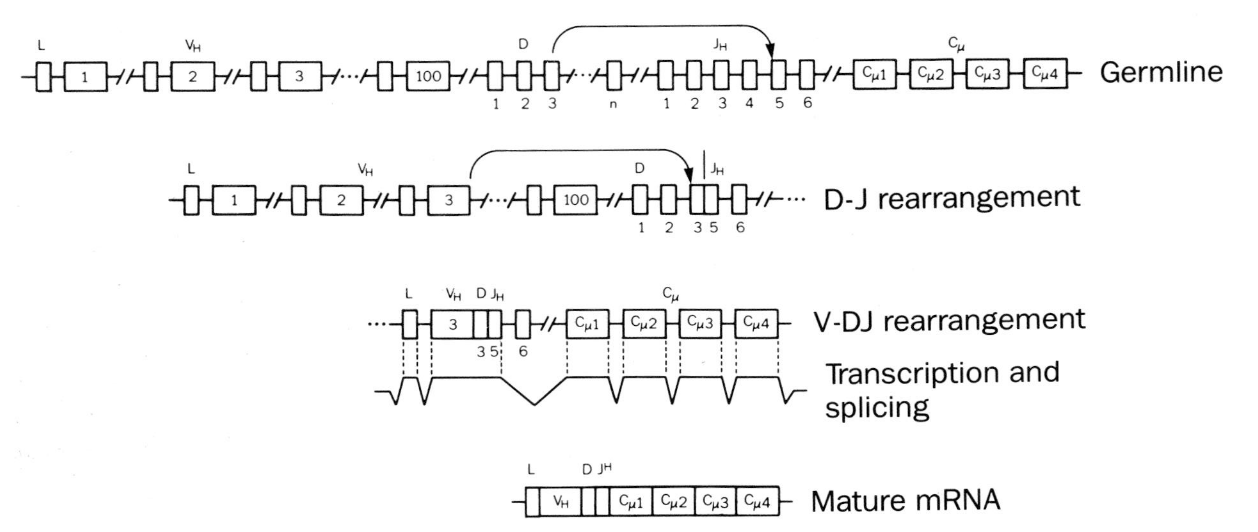

Figure 5.3. *Organization, rearrangement, and transcription of human immunoglobulin heavy chain genes.* The heavy chain locus contains about 100 V_H genes, approximately 30 D, six J_H, and 11 C_H genes, only $C\mu$, the most 5′, is shown. The first DNA rearrangement step joins D to J_H, the second brings V_H to DJ_H. As with light chains, introns are spliced out of primary transcripts to give mature heavy chain mRNA.

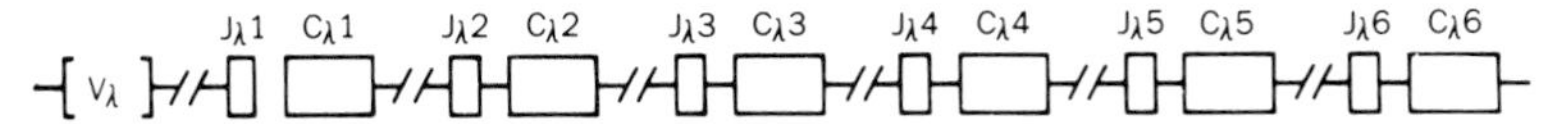

Figure 5.4. *Organization of human immunoglobulin λ chain genes.* An undetermined number of V_λ genes are located 5′ to six J_λ and six C_λ genes. In contrast to heavy chain and κ light chain loci, each J_λ gene is associated with one C_λ gene. As with κ light chains, a single DNA rearrangement event joints a V_λ gene to J_λ. Again, introns are spliced out to yield mature mRNA.

are subsequently spliced out, leaving only the exon sequences. Thus, introns are not translated into protein. The first V gene exon encodes most of the *leader sequence* or *signal peptide* analogous to those found in many secreted proteins. This is the first part of the protein synthesized (the amino terminus) and this peptide interacts with the endoplasmic reticulum, important in the intracellular transport and processing of the protein. The signal peptide is removed from the antibody before it appears in the membrane or is secreted. The second V gene exon encodes the majority of the V region, about 95 amino acids. D genes are variable in length and may encode 1–15 amino acids. J genes encode 15–20 amino acids.

In heavy chain gene rearrangement (Figure 5.3), the 3′ end of a D gene is brought to the 5′ end of a J_H gene, with loss of the DNA (often including other D and J_H genes) that used to separate them. This is called *D-J rearrangement*. Next, the 3′ end of a V_H gene is juxtaposed to DJ, again with loss of intervening DNA (*V-DJ rearrangement*). Light chains undergo only one rearrangement event, V to J (Figure 5.2).

V, D, and J genes have flanking sequences which are important in the mechanism of rearrangement. These are the *heptamer* (seven base pairs, or bp) and *nonamer* (nine bp) sequences which are separated by a *spacer* of either 12 or 23 bp (Figure 5.5). A gene with a flanking sequence containing a 12 bp spacer may only join to a gene whose flanking sequence has a 23 bp spacer. This is called the *12–23 base pair rule*. Presumably, these sequences are recognized by DNA-binding proteins focusing the activity of endonucleases and ligases performing the splicing.

Immunoglobulin genes are transcribed at a low rate even before they rearrange. Very little protein (or none) is synthesized from these RNAs, and they are called *sterile transcripts*. After rearrangement, the *immunoglobulin enhancers* become active. An enhancer is a nucleotide sequence which can increase the rate of transcription of genes near it. The heavy chain immunoglobulin enhancer is situated in the intron between the J_H and C_H genes. Similarly, the κ enhancer resides in the intron

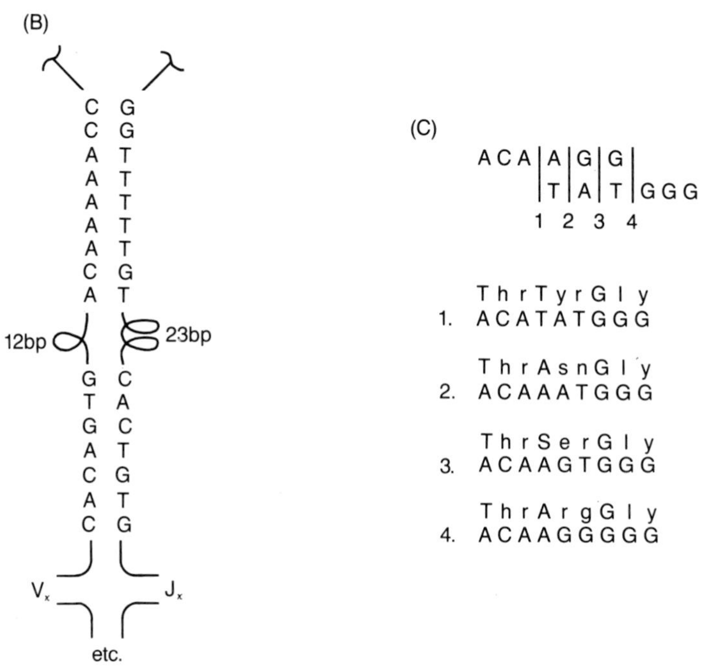

Figure 5.5. *Some details of immunoglobulin gene rearrangement.*

A. *The heptamer and nonamer recombination sequences.* The 3′ flanking sequences of V_κ genes have the palindromic heptamer recombination signal sequence, a 12 bp spacer, and the AT-rich nonamer signal sequence. The 5′ flanking region of J_κ contains the sequences in the reverse order, and separated by a 23 bp spacer. The immunoglobulin gene recombination mechanism only operates when one flanking sequences has a 12 bp spacer, and the other a 23 bp spacer (the *12–23 bp rule*). The flanking sequences of λ gene segments are identical, but transposed. The 12 bp spacer is 5′ to J_λ, and the 23 bp spacer 3′ to V_λ. V_H genes have 3′ 23 bp spacers, D genes are flanked both 5′ and 3′ with 12 bp spacers, and J_H genes have 5′ 23 bp spacers.

B. *A possible Ig gene rearrangement intermediate.* In one proposed model of gene rearrangement, the flanking DNA separates into single strands permitting annealing of signal sequences from different flanking regions, with "looping out" of the spacers and the intervening DNA. This intermediate DNA structure is, presumably, stabilized by (a) specialized binding protein(s) while endonucleases and ligases perform the actual splicing.

C. *Consequences of imprecise gene segment joining.* Ig gene segments need not always be spliced together at the same positions within their coding sequences. As long as the reading frame is maintained, a potentially functional protein may be generated. In the example here, four different in-frame splices are possible. Splicing at each position creates a codon for a different amino acid at the union of the two gene segments.

separating J_κ from C_κ. V gene rearrangement brings a promoter near the enhancer region, and immunoglobulin gene transcription increases greatly.

Antibody diversity

How can an immune system effectively cope with the tremendous variety of antigens in nature without requiring a large fraction of the genome to encode its antigen receptors? The answer lies in the very efficient usage of DNA displayed in the gene organization and rearrangement process described above.

Consider the number of different heavy chain VDJ combinations which may be assembled. Estimates of the number of V_H genes range from 100–1000. For the purposes of this calculation, let us assume the number is 100. There are about thirty D, and six J_H genes. Any V_H can join to any D, and any D to any J_H, thus we may assemble 18,000 different heavy chain VDJ units. We have amplified the germline-encoded diversity by 180-fold, simply by combining V_H gene segments with various D and J_H segments.

Additional mechanisms increase the number of VDJ genes which may be generated. First, the joining of D to J_H and V_H to DJ is not precise; it need not always occur at the same position in the gene segments. This results in different codons at the joints, and also in different D region lengths (Figure 5.5). Second, analysis of nucleotide sequences of rearranged immunoglobulin genes has revealed that at the gene segment junctions, one may find nucleotides which are not encoded by any identifiable genomic DNA. These nucleotides are believed to be inserted at the gene joining points by the enzyme *terminal deoxyribonucleotidyl transferase*, and are sometimes designated the *N region*. It is difficult to precisely quantify the additional variety generated by imprecise joining and N regions, but let us conservatively estimate it as a ten-fold increase, yielding 180,000 different possible VDJ genes in our hypothetical calculation. A similar calculation for the κ locus (assuming 200 V_κ genes) gives about 4,000 possible κ VJ genes, assuming a four-fold increase in diversity due to imprecise V_κ–J_κ joining. If any heavy chain can pair with any light chain (not proven, though often assumed) this results in 7.2×10^8 possible V_H–V_κ pairs. To this number is added the number of possible V_H–V_λ pairs.

Recent analysis of nucleotide sequences of rearranged murine V_H genes strongly suggests that one D gene segment may join to another

D during rearrangement. Clearly, this would generate additional diversity from genomic sequences.

One additional mechanism increasing combining site diversity exists: *somatic mutation.* At some time after rearrangement is complete, the heavy and light chain V genes may accumulate point mutations. One gene may collect up to 10 or more nucleotide changes, some fraction of which result in an amino acid substitution. These changes may alter antibody affinity, or may even alter specificity.

It should be mentioned that all of these mechanisms which create antibody diversity also create some cellular wastage. A series of gene rearrangements may create non-functional genes by altering the reading frame or producing an early termination codon. In addition, some proportion of V genes are pseudogenes and encode non-functional proteins. In order to produce an antibody, two series of rearrangements are required, one for heavy and one for light chains. Each series of rearrangements is susceptible to error. The fraction of B cells lost to aberrant Ig gene rearrangements is unknown, but it is probably greater than 50%. In summary, four mechanisms generate combining site diversity: multiple V, D, and J genes (germline-encoded diversity); rearrangement with imprecise joining of gene segments; pairing of heavy and light chains (combinatorial diversity); and somatic mutation. These mechanisms allow for efficient use of a relatively small portion of the genome to provide a sufficiently large *repertoire* of antibody combining sites, a population of antibody specificities so diverse that no matter what epitopes are encountered, a complementary antibody may be found.

IMMUNOGLOBULIN PROTEINS

The four chain unit

The fundamental architectural theme of an immunoglobulin molecule is a four chain structure containing two identical light (L) chains of $M_r = 23{,}000$, and two identical heavy (H) chains of $M_r = 55{,}000$ (Figure 5.6).

The H and L polypeptides assume characteristic secondary and tertiary structures forming the *immunoglobulin domains* (Figure 5.7). Light chains have two domains; heavy chains have four or five. The domain structures formed by an individual H or L chain, and the complete four chain unit, H_2L_2, are maintained by covalent and

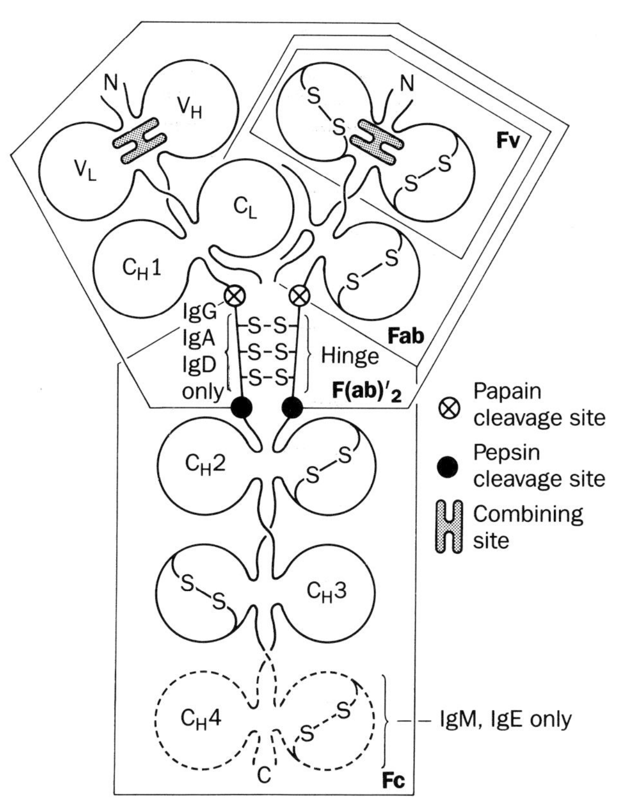

Figure 5.6. *Structure of an antibody molecule.* The basic four-chain unit is diagrammed, along with some of the fragments that may be generated by limited proteolysis. Each circle represents an immunoglobulin domain, intra-domain and inter-chain disulfide bonds are indicated (S-S). An additional disulfide bond joins the heavy chain to the light chain (not shown). The only exception to this is the A2m(1) allotype of the human IgA2 subclass in which light chains are disulfide bonded to one another, not the heavy chain. Only IgG, IgA, and IgD classes have distinct hinge regions; only IgM and IgE have four C_H domains. N and C indicate amino and carboxyl polypeptide termini, respectively.

non-covalent interactions. A domain consists of approximately 110 amino acids organized into two regions of β-pleated sheet held apposed (sandwich fashion) by a disulfide bridge. The four chain unit is also usually stabilized by disulfide bonds between H and L chains, and between the two H chains (Figure 5.6). Non-covalent interactions stabilizing antibody structure are hydrogen bonds, electrostatic forces, van der Waal's forces, and hydrophobic interactions (see below).

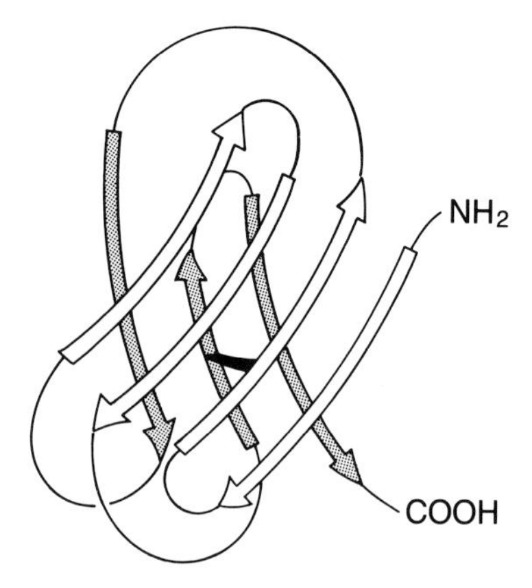

Figure 5.7. *The immunoglobulin domain.* Antibodies contain predominantly β-pleated sheet secondary structure. The strands comprising the sheets (arrows) are joined by short peptide segments. The black band indicates the intra-domain disulfide bond. (Adapted from Edmundsen et al., 1975.)

The V_L region (the N-terminal half of the light chain) is approximately 95 amino acids long; the J_L region contains 13–15 amino acids (Figure 5.8). The C_L region is approximately 110 amino acids long. The V_H region is comprised of the 100 amino terminal residues of the heavy chain. Proceeding toward the C terminus, the next one to 15 residues constitute the D region, followed by the 15–20 residues of the J_H region (Figure 5.7). The V_H, V_L, and C_L regions all consist of only one domain, while C_H contains three or four domains. The domains of the C_H region are numbered with C_H1 nearest V_H, and C_H3 or C_H4 at the carboxyl terminus (Figure 5.6).

The combining site

As mentioned above, the variable regions of the H and L chains together form the antibody *combining site*, or *paratope*, the portion of the antibody interacting with antigens (epitopes). Since one H chain and one L chain together form a paratope, a single four chain unit contains two combining sites, i.e., it is *bivalent*.

If a group of H or L chain V region amino acid sequences are compared to one another, a greater variety of amino acids occur in certain positions in the sequence in relation to others. These highly variable positions are clustered in the V region sequence, and are

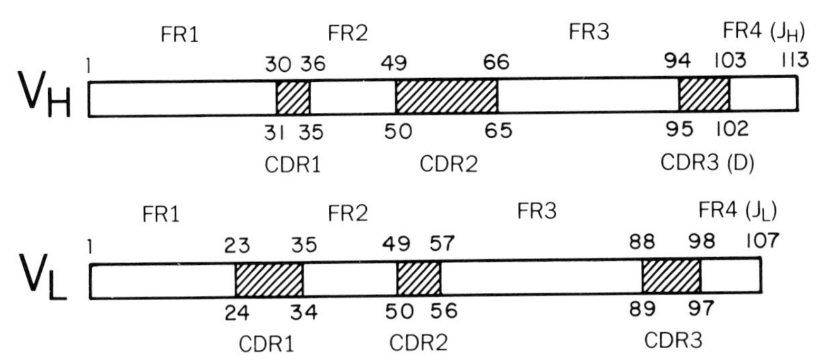

Figure 5.8. *Subdivisions of the variable region.* Immunoglobulin V regions are subdivided into four *framework regions* (*FR*) with three *complementarity-determining regions* (CDR) between them. A V_H gene encodes through the end of FR3. CDR3 is encoded by a D gene, and FR4 by a J_H gene. V_L genes encode almost to the end of CDR3, the remainder and FR4 is encoded by a J_L gene. Not all V_H or V_L genes are the same length. V genes are grouped into families based on their sequence similarities and differences (see text). Genes of a particular family may have a longer CDR2, a shorter FR1, etc. The numbering scheme shown here was devised by Kabat et al. (1987) in order to have a standardized system for comparing Ig gene sequences. The reader is referred to that text for additional details of the numbering system.

designated the *hypervariable regions.* Intervening between them are the less variable *framework regions.* It is the hypervariable regions which make contact with antigen. For this reason, they are frequently referred to as *complementarity-determining regions* (*CDR*'s, Figure 5.8).

The hinge region

Between C_H1 and C_H2 of certain antibody classes (IgG, IgA, and IgD, see below) lies the *hinge region* (Figure 5.6). This span of 12–15 amino acids confers additional flexibility to these antibodies, allowing a single four chain unit to bind to two (identical) epitopes on one antigen particle or molecule. The hinge region contains most of the disulfide bonds linking the two H chains, and its structure is characteristic of a particular antibody class or subclass. IgM and IgE possess hinge-like regions at the C terminal end of C_H2.

Carbohydrate

Immunoglobulins contain 3–12% sugar, by mass. Associated predominantly with the heavy chain, the carbohydrate moieties are

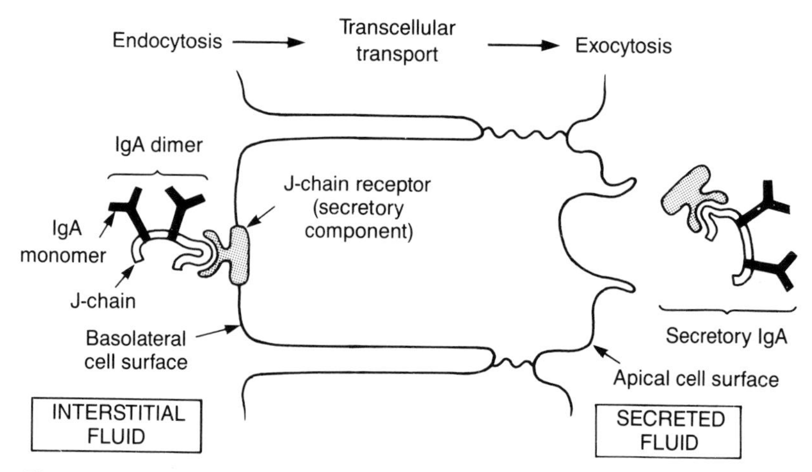

Figure 5.10. *Secretion of IgA.* An IgA dimer consists of two IgA molecules held together by the same J chain as in pentameric IgM (Figure 5.9). The basolateral surfaces of secretory epithelial cells possess receptors for the J chain in polymeric Ig (IgA or IgM). Binding to this receptor induces internalization of the poly-Ig-receptor complex. The entire complex is transported to the cell's apical surface where it is secreted. The "receptor" is now called the *secretory component* (*SC*).

surface (Figure 5.10). Some pentameric IgM is also secreted in this fashion.

IgD

IgD is present at low levels in the serum, < 0.1 mg/ml. Along with IgM, however, IgD is a predominant class among the surface receptors of mature unstimulated B cells. IgD has $M_r = 170{,}000$, contains about 12% carbohydrate, and has a serum half-life of three days.

IgE

IgE has the lowest concentration in normal human serum, < 0.001 mg/ml, and has $M_r = 190{,}000$, consisting of 12% carbohydrate. IgE also has a short half-life of about three days. IgE plays a prominent role in allergic reactions (see Chapter 10).

IgD, IgG, and IgA have three constant region domains; IgM and IgE have four (Figure 5.6). The fundamental properties of the different antibody classes and subclasses are summarized in Table 5.II.

Table 5.II. CHARACTERISTICS OF ANTIBODY CLASSES AND SUBCLASSES

Class Subclass	M_r	No. C_H domains	Serum concen-tration (mg/dl)	Serum half-life (days)	Properties
IgM	950,00 (pentamer)	4	150	5	Complement fixation, predominant in primary response
IgG	150,000 (monomer)	3	1350		Complement fixation, opsonization, placental transfer, predominant in secondary response
IgG1			900	23	
IgG2			300	23	
IgG3			100	8	
IgG4			50	23	
IgA	400,000 (dimer)	3	200	6	Predominant in secretions
IgA1					
IgA2					
IgD	170,000 (monomer)	3	5	3	B cell receptor
IgE	190,000 (monomer)	4	0.1	3	Immediate hypersensitivity

Proteolysis of antibodies

The hinge region and areas near it, due to their loose folding and exposure to solvent, are susceptible to proteases. The sulfhydryl protease *papain* cleaves IgG into three fragments. Two of these fragments are identical, and each bears one combining site. Since they react with antigen, these fragments are called *Fab* (*antigen-binding*). An Fab fragment contains a complete L chain, and the V_H and C_H1 domains of the H chain (Figure 5.6). The other papain cleavage product spontaneously *crystallizes* in cold, neutral solution and is called *Fc*. This fragment contains the C_H2 and C_H3 domains of the IgG H chains. Papain cleaves in the hinge region.

The acid protease pepsin cuts the H chain at a point just to the N terminal side of C_H2. As a result, the hinge regions remain joined, yielding a fragment containing both paratopes. This fragment is called $F(ab)'_2$. Since $F(ab)'_2$ has two combining sites, it may bind two particles simultaneously, and agglutinate and precipitate antigens (see below). Fab has only one combining site and cannot participate in these reactions.

Upon chemical reduction, Fab fragments may be separated into intact light chains, and the *Fd* fragment consisting of V_H and C_H1 domains. Further proteolysis of Fab yields *Fv*, the isolated V regions of the heavy and light chains with an intact combining site. Similarly, Fc can be split to give isolated C_H2 and C_H3 domains.

Antigenic determinants of immunoglobulins

Under appropriate circumstances, antibodies elicit specific immune responses, as might any other globular protein. Immunoglobulin antigenic determinants are defined by their location on the antibody molecule, and their occurrence on antibodies of different classes, from different individuals, or different species. The three main categories are *isotypic*, *allotypic*, and *idiotypic determinants.*

Isotypic determinants

The term *isotype* is often used interchangeably with the terms *class* and *subclass.* Isotypic determinants, then, are immunoglobulin epitopes characteristic of a class or subclass of antibody heavy or light chains. Antibody isotypes are defined in relation to heavy and light chain *constant regions only.* All isotypes are found among the immunoglobulins of all healthy individuals of a species. Thus, antibodies specific for isotypes are easily obtained only through immunization of a different species (*heterogeneic* immunization).

Allotypic determinants

Some of the genes encoding antibody constant regions are *polymorphic.* That is, there exist several different alleles for the genes encoding certain Ig isotypes. The proteins encoded by different isotype alleles are very similar, but not identical. The epitopes distinguishing a particular antibody encoded by allele "a," from an antibody of the same isotype encoded by allele "b", are called allotypic determinants (*allotypes*).

Human allotypes were originally discovered in studies of a particular type of antibody known as a *rheumatoid factor* (*RF*). RFs are IgM antibodies which bind a particular IgG subclass. However, an RF from one individual does not bind to all human IgGs of a given subclass. RFs only react with antibodies encoded by a particular allele within a subclass. Thus, RFs define allotypes. Allotypes have been defined for all subclasses of IgG, IgA2, and κ light chains.

Table 5.III. KM ALLOTYPES

Allotype	C_κ amino acid number: 153	191
Km (1)	Val	Leu
Km (1,2)	Ala	Leu
Km (3)	Ala	Val

In most cases, protein structural data correlate allotypes with amino acid changes between them. Table 5.III shows the amino acids distinguishing human Km allotypes. Human allotypes are named with the antibody class or subclass, followed by the letter "m" and a number in parentheses, for example, Km(1), G1m(1), G3m(5), A2m(1), etc.

Allotype-specific antibodies may be obtained by immunizing an individual of the same species who does not express the allotype (*allogeneic* immunization). Allotypes are valuable in anthropological studies of the genetic relatedness of different populations. In some countries, allotypes are accepted as legal evidence in paternity disputes.

Idiotypic determinants

The variety of idiotypic determinants, or *idiotypes* occurring in one individual is vastly greater than the number of isotypes or allotypes. This is because idiotypes are determined by antibody V regions, and there are many more V region genes than there are C region genes. Idiotypes may be unique to one antibody, may be shared by antibodies with the same or different specificities, in the same or different individuals of a species, and may even be found on antibodies from different species. Within a species, idiotypes may be associated with any isotype, but often segregate with allotypes. This is not surprising since V and C genes are closely linked on the same chromosome. Since antibody specificity is a property of the V region, and isotypes and allotypes are associated with C regions, a given isotype or allotype may occur in association with any specificity or idiotype which the individual may produce.

Recall that one of the mechanisms generating antibody diversity is somatic mutation. After B cell maturation, rearranged V genes may accumulate point mutations. Since somatic mutation operates on variable regions, isotypic and allotypic determinants are not affected.

On the other hand, antibody affinity and even specificity, as well as idiotype expression, may be altered by somatic mutation.

It is possible to obtain antibodies specific for an idiotype (*anti-Id antibodies*) in the same individual that produced the Id-bearing antibody. For example, pregnant women produce antibodies which bind the major histocompatibility (MHC) proteins of the fetus which were derived from the father (foreign to the mother). At the same time, the mother also produces anti-idiotype antibodies which bind to her own anti-MHC antibodies. The production of idiotypes and anti-idiotypes in a single individual forms the basis for a theory of immune system regulation known as the *idiotype network theory* (discussed in Chapter 6).

BIOLOGICAL PROPERTIES OF IMMUNOGLOBULINS

The smallest fully functional antibody molecule is a single four chain unit. Since it is formed of two identical L chains and two identical H chains (H_2L_2), the molecule is bivalent in all surface characteristics: combining sites, antigenic determinants, and heavy chain C region structures interacting with various cellular receptors (Figure 5.11).

V region functions

The antibody-antigen interaction

If the antibody paratope, in isolation, may be said to have a function, it is its specific interaction with a complementary epitope. The combining site-antigen interaction is determined by spatial arrangements of chemical functional groups, much as are the interactions of enzymes with their substrates. To a large extent, epitope and paratope define one another, although subtly different paratopes may interact with the same epitope with different affinities, and vice-versa (see Chapter 3).

Four types of physical/chemical interactions occur between paratopes and epitopes. *Van der Waal's forces*, or *induced dipole interactions* are extremely short-range attractions occurring between atoms without permanent dipole moments. Electron clouds of neighboring atoms fluctuate producing mutually attractive temporary dipoles. *Hydrogen bonds* are highly directional forces occurring when hydrogen atoms interpose between two electronegative atoms such as oxygen (Figure 5.12).

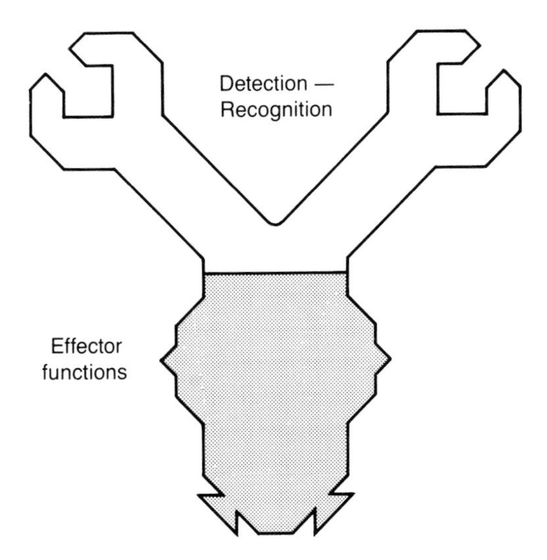

Figure 5.11. *Categories of antibody function.* Antibody molecules possess two types of function: the binding of antigen (detection/recognition) and several biological activities (effector functions). Detection is a function of the variable region, while effector functions are determined by the C region. Antigen binding causes a conformational change in the C region which is important for some antibody functions (for example, complement fixation, and interaction with some Fc receptors). Note that antibodies are bivalent for antigen binding. As will be seen in this and the following chapter, this property is very important for antibody functions in immune responses.

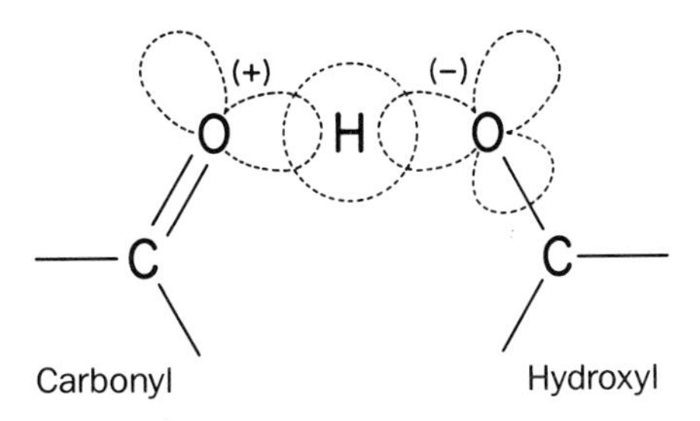

Figure 5.12. *Diagram of a hydrogen bond.* In this example, a hydrogen bond is formed between alcohol and carbonyl groups. The highly electronegative alcohol oxygen leaves the hydrogen nucleus (proton) somewhat exposed, available to interact with an electron pair of the nearby carbonyl oxygen. The carbonyl acquires a partial positive charge, the alcohol a partial negative charge. The strength of the hydrogen bond is greatest when all three participating nuclei are colinear. Other elements forming hydrogen bonds are nitrogen and fluorine.

Electrostatic forces are important when functional groups in the combining site and on the antigen bear electrostatic charges or permanent dipole moments. In these instances, the strength of the interaction may depend on pH. So-called *hydrophobic interactions* refer to the increased entropy resulting when water is excluded from regions of the antibody-antigen interface containing aliphatic or aromatic residues. This is the same principle by which "oil and water don't mix". Examples of these chemical interactions are shown in Figure 5.13.

We often visualize antibody combining sites as "grasping" since this is a comfortable conceptualization of a relatively stable noncovalent chemical interaction. This "lock and key" idea is also rooted in notions of enzyme and substrate complementarity which has many parallels with antibody-antigen interactions. However, not all epitopes conveniently

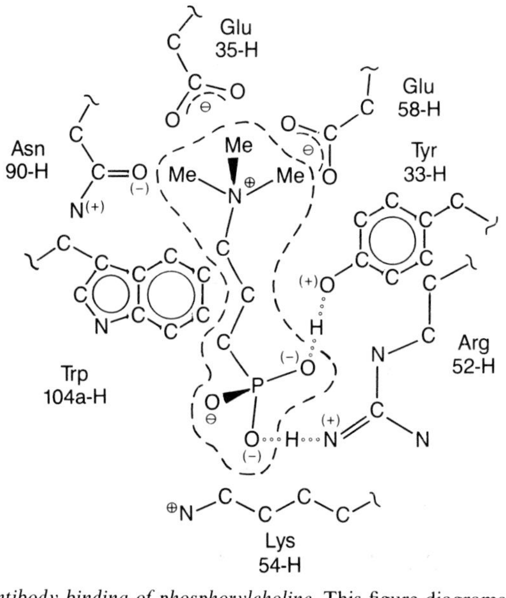

Figure 5.13. *Antibody binding of phosphorylcholine.* This figure diagrams the interaction of the hapten phosphorylcholine (PC, enclosed with the dotted line) with the residues of an antibody combining site. Three different types of interactions are depicted. Asn-90, Glu-35, and Glu-58 all bear full or partial negative charges and attract the positively charged quaternary ammonium group of PC. Similarly, the ε-amino group of Lys-54 is positively charged and attracts the negatively charged phosphate moiety of PC. Tyr-33 and Arg-52 participate in hydrogen bonds with phosphate oxygen atoms. Trp-104a has a hydrophobic interaction with the aliphatic backbone of PC. All of the interacting residues shown are contributed by the antibody heavy chain. (Adapted from Capra and Edmundson, 1977.)

protrude from an antigen's surface so that they may be surrounded by CDR amino acid side chains as depicted in Figure 5.13. It is now also clear that paratope structures are not limited to concavities that "hold" epitopes. Combining sites may have virtually any three-dimensional shape. Consider the conformational epitope of hen egg lysozyme (HEL) described in Chapter 3 (Table 3.I). The CDR residues of a HEL-specific antibody making contact with this epitope are listed in Table 5.IV. In this case, the antigen-antibody interaction occurs over a planar surface approximately 750 Å^2 in area.

For the reaction of antibody (Ab) with antigen (Ag) as written in equation (1) (where Ab · Ag denotes the antibody-antigen complex), the *association constant* for the reaction (K_a) is given by equation (2), where [X] denotes the molar concentration of X.

$$\mathrm{Ab} + \mathrm{Ag} \rightleftharpoons \mathrm{Ab} \cdot \mathrm{Ag} \qquad (1)$$

$$K_a = \frac{[\mathrm{Ab} \cdot \mathrm{Ag}]}{[\mathrm{Ab}][\mathrm{Ag}]} \qquad (2)$$

The association constant is often called the *affinity constant*, or simply the affinity of antibody for antigen. Antibody affinities are most often in the range 10^5–10^{11} L/mole. Antibody affinity is biologically significant since it determines the minimum concentrations of reactants at which effective interactions occur. High-affinity antibodies interact with their complementary antigens when the concentration of either is low.

Table 5.IV. RESIDUES OF ANTIBODY D1.3 WHICH CONTACT AN EPITOPE OF HEN EGG LYSOZYME

#	Heavy chain Residue	CDR	#	Light chain Residue	CDR
30	Thr	1	30	His	1
31	Gly	1	32	Tyr	1
32	Tyr	1	49	Tyr	2
52	Trp	2	50	Tyr	2
53	Gly	2	91	Phe	3
54	Asp	2	92	Trp	3
99	Arg	3	93	Ser	3
100	Asp	3			
101	Tyr	3			
102	Arg	3			

See Figure 5.7 for the locations of the heavy and light chain complementarity determining regions (CDRs).
(Data from Bentley et al., 1989.)

Precipitation

When complementary antibodies and antigens are mixed in solution, complexes form and precipitate. The amount of precipitate is related to antigen and antibody valence, reactant concentrations, and antibody affinity, which also may depend on the pH and ionic strength of the solution. Note that precipitation requires antibody and antigen valences greater than one. An Fab fragment cannot precipitate antigen. Similarly, a monovalent antigen cannot interact with more than one antibody. Since secreted IgM molecules are pentameric (10 combining sites) they are very good precipitators. Figure 5.14 shows a typical *precipitin curve*, one method of measuring the reaction of antibody and antigen.

When added to antibody-antigen mixtures, large polysaccharides and other high molecular weight charged compounds enhance precipitation, probably by reducing the amount of water available for solvation of antigen and antibody. The same effect is obtained when precipitation is carried out in a semi-solid medium, such as an agar gel. When one juxtaposes two agar mixtures, one containing antibody and the other antigen, a visible zone of precipitate forms at the interface where the two components diffuse together. This is known as *immunodiffusion*. In a variation of this technique, agar is poured into a petri dish, and small wells are made. Antigen and antibody are placed in separate wells. If antigen-antibody reaction occurs, a line of precipitate forms where the reactants meet after diffusing into the gel. This is called an *Outchercherlony double diffusion reaction*, named for the scientist who developed the method. By examining the patterns of precipitation in these reactions, one may determine whether two antigens share determinants, or whether two antibody preparations share specificities (Figure 5.15).

Another variant of this technique is *immunoelectrophoresis*. In this method, electrophoresis in one dimension is coupled with immunodiffusion in a perpendicular direction. With this technique, one may analyze much more complex mixtures of antigens and antibodies than with simple immunodiffusion (Figure 5.16).

Two additional methods related to precipitation of antibody-antigen mixtures have revolutionized the study of antigen-antibody interactions and clinical chemistry. These are the *radioimmunoassay* (*RIA*), and the *enzyme-linked immunosorbent assay* (*ELISA*).

The radioimmunoassay is a solution phase immunoprecipitation in which one reactant is radioactively labelled. Following the reaction, precipitate is removed by ultracentrifugation. Either the quantity of radioactivity precipitated, or remaining in solution is measured. This

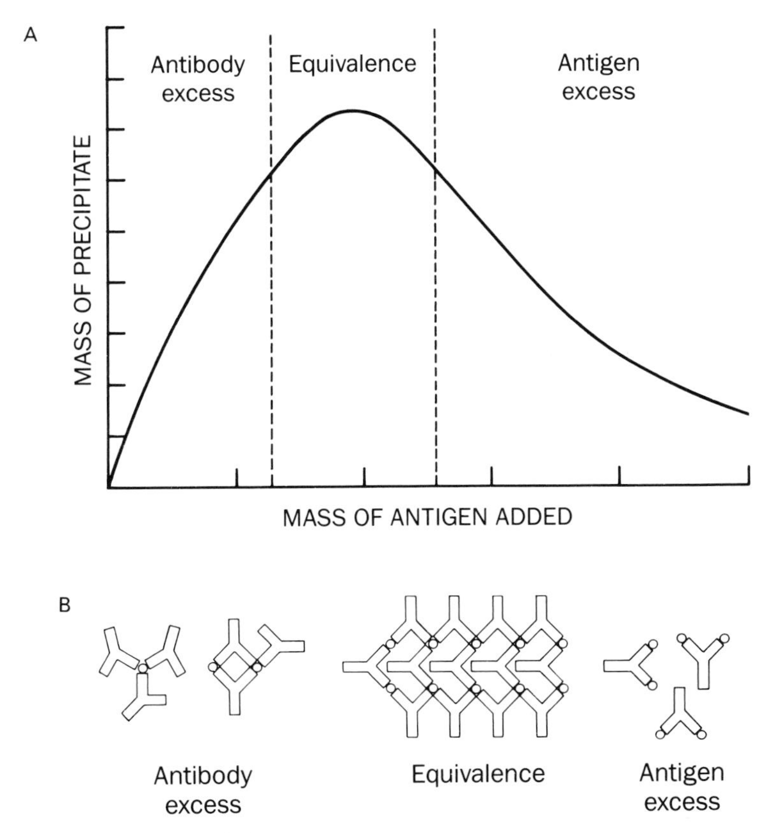

Figure 5.14. *Antigen-antibody precipitation.*

A. *A precipitin curve.* Increasing amounts of antigen are added to a series of tubes containing a fixed amount of antibody. The fraction of antigen precipitated remains near 100% until the number of antigenic determinants is roughly equal to the number of combining sites (equivalence). Progressively less precipitate is formed as the amount of antigen is increased.

B. *Antigen-antibody complexes.* Small complexes are formed in antibody excess, several antibodies binding to each molecule of antigen. At equivalence, very large antibody-antigen lattices may form. In antigen excess, combining sites are quickly saturated and few large complexes form.

technique is frequently applied to determine the presence or absence of a particular antigen or antibody specificity in a solution (e.g., human serum). RIA is very sensitive and can detect nanogram quantities of antigen or antibody.

A variant of this technique, *solid phase RIA* (*SPRIA*), is widespread in research and clinical chemistry. In this method, a solution containing

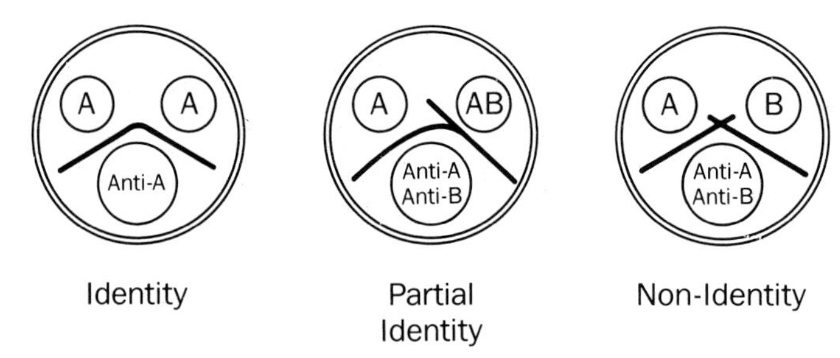

Figure 5.15. *Outcherlony double diffusion.* Antigens or mixtures are placed in the two upper wells of the agar plate, antiserum in the lower. *Left.* In this case an antiserum reacts with an antigen present in both wells; a continuous line of precipitation forms: the line of identity. *Center.* Two different antigen-antibody interactions produce two lines of precipitation. Here, one antigen (A) is present in both wells producing a line of identity, while antigen B forms another line of precipitation. Depending on the diffusion of the different antigens, the lines of precipitation may be partly continuous, or may intersect. *Right.* No specificities are shared between antigen wells; two intersecting lines of precipitation are formed.

one reactant is placed in a well in a dish made of polystyrene, or polyvinylchloride. These plastics will (more or less) irreversibly bind most proteins and many polysaccharides. Subsequently, the plate is washed, and the radioactively labelled reactant added. After washing again, the amount of bound radioactivity is measured. This is a *direct*

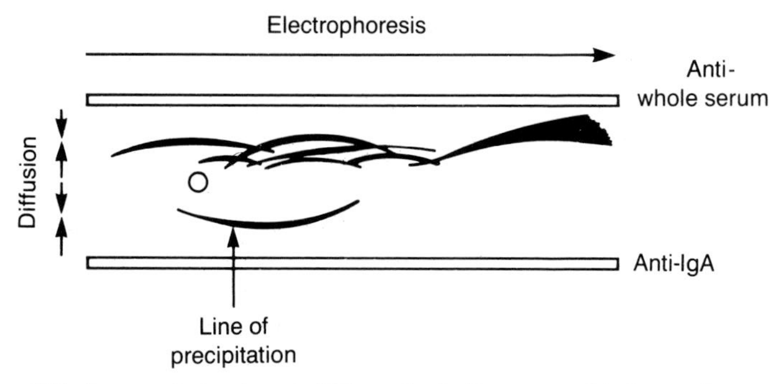

Figure 5.16. *Immunoelectrophoresis.* This method often uses a thin layer of agarose gel poured onto a microscope slide. A small amount of the solution to be tested (e.g., serum) is placed in a small well in the gel. Electrophoresis is carried out along the long axis of the slide. Next, antiserum, or a solution of antibodies, is placed into troughs cut along the sides of the gel. Diffusion (immunodiffusion) then occurs along the short axis of the slide. Electrophoresis prior to immunodiffusion permits much better resolution of complex antigen mixtures. In this diagram, a drop of normal serum is electrophoresed and reacted with a mixture of antibodies to all serum components (upper trough), and a reagent specific for IgA (lower trough).

binding assay. The labelled component may be either antigen or antibody. The reader may easily envision more complex strategies involving more than two components (*indirect binding assay*, Figure 5.17). SPRIA finds wide and ever-increasing application in *immunodiagnosis* in clinical laboratories.

ELISA is very similar to SPRIA except that one labels a reagent with an enzyme instead of with a radioactive isotope (Figure 5.17). The enzyme is quantified by measuring the amount of chromophore generated from a chromogenic substrate in a given period of time. One enzyme-substrate system often used is alkaline phosphatase and *p*-nitrophenyl phosphate. ELISA is also widely applied in clinical chemistry.

These techniques have also been combined with histological methods, yielding more large words such as *immunohistochemistry*. Antibodies labelled with an enzyme (e.g., horseradish peroxidase) produce pigment deposits when incubated with an appropriate substrate (e.g., diamino-benzidine). Radioactively labelled antibodies are detected when the

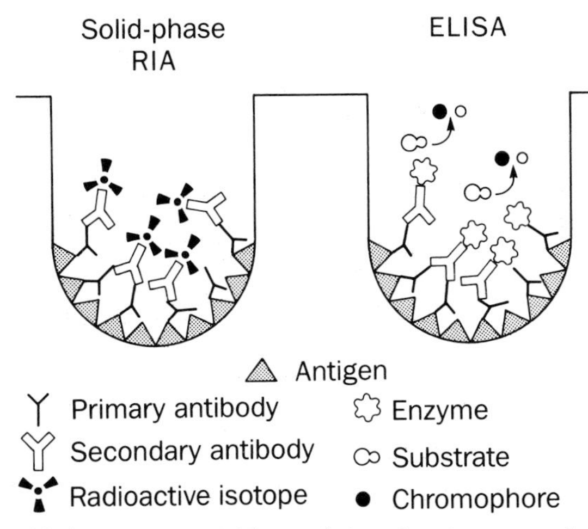

Figure 5.17. *Solid-phase immunoassay*. These techniques have in common the immobilization of one reactant on a plastic surface. All subsequent reactions take place on this surface.

A. *Solid-phase radioimmunoassay (SPRIA)*. The experimental design depicted here is only one of many possible. Plates are coated with an antigen. The primary antibody (e.g., human antiserum) is allowed to react. Binding of the antiserum is detected by a radiolabelled secondary antibody (e.g., goat anti-human immunoglobulin).

B. *Enzyme-linked immunosorbent assay (ELISA)*. The experimental design is identical to that in A except the secondary reagent is coupled to an enzyme rather than a radioactive isotope. The readout of the assay is the optical density of a chromophore generated by enzyme activity.

specimen is coated with a photographic emulsion; the radioactive emissions create silver grain deposits (see Figure 6.3). Antibodies labelled with fluorescent compounds (e.g., fluorescine, rhodamine) may be visualized in a fluorescence microscope (see Figure 2.10). Antibodies labelled with gold can be detected in electron micrographs (see Figure 8.4). All of these methods are powerful tools for localizing particular molecules within cells and tissues.

Agglutination

Precipitation and agglutination are conceptually identical. The difference between them is that while precipitation involves soluble antigens and antibodies, agglutination denotes the formation of complexes of antibodies with relatively large particles such as bacteria, or erythrocytes (Figure 5.18). As does precipitation, agglutination depends on antibody multivalence. Not surprisingly, the same factors influencing precipitation (antibody concentration, pH, ionic strength, affinity, etc.) also affect agglutination. Pentameric IgM is a good agglutinator, as well as a good precipitator. In addition, the density of antigenic determinants on the particle surface, as well as electrostatic interaction among the particles, will affect agglutination. For example, most cells have a net negative surface charge (the zeta potential) and tend to repel one another.

Agglutination tests are often used to detect serum antibodies with a particular specificity. The specificity may be one normally present on

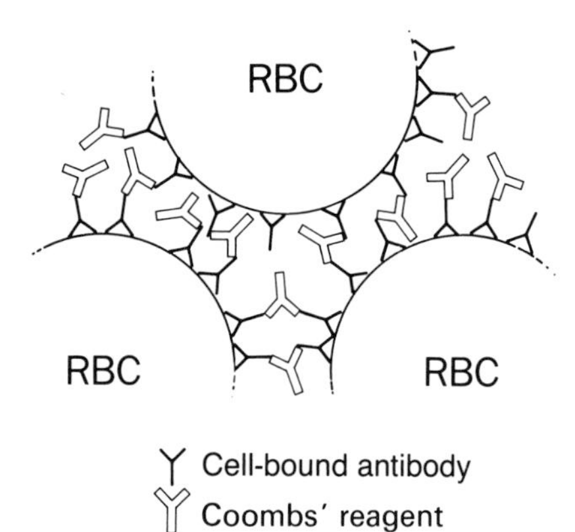

Figure 5.18. *Coombs' test.* A patient's red blood cells (RBC) are incubated with an anti-human immunoglobulin reagent (Coombs' reagent). The Coombs' reagent agglutinates erythrocytes if they have antibodies bound to their surface.

the target cell (usually an erythrocyte), or it may not, in which case it is artificially coupled to the red cell. Serial dilutions of serum are mixed with target cells, and antibody levels are quantified as the highest dilution causing agglutination. In sera with high titers of specific antibody, agglutination may not be observed until some dilution is made. Lower dilutions not causing agglutination are called the *prozone*. At low dilutions, target epitopes are rapidly saturated with antibody, preventing binding of one antibody to two particles.

A common clinical agglutination test is the *Coombs' test*. This test detects antibodies on erythrocytes. These antibodies (*Coombs' antibodies*) alone are unable to agglutinate red cells. A positive Coombs' test results when anti-human globulin (*Coombs' reagent*) binds antibodies coating the erythrocytes and causes agglutination (Figure 5.18). The *indirect Coombs' test* detects erythrocyte-reactive antibodies in serum. In this test, erythrocytes are incubated with a patient's serum, washed, then exposed to Coombs' reagent.

Neutralization

When an antibody binds to a virus in such a way that it is no longer infectious, the virus is said to be *neutralized*, and such an antibody is called *neutralizing*. The same terminology is used when a toxin is rendered biologically inactive by antibody binding. Many bacteria, parasites, and animals produce toxins with a variety of detrimental biological effects. One of the most deadly, on a molar basis, is botulinus toxin, a neurotoxin produced by the bacterium *Clostridium botulinum*. The only hope for survival in botulism poisoning is the timely injection of a large amount of antibodies binding the toxin (*antitoxin*, usually prepared in horses). Similarly, when an individual has been bitten or stung by a venomous animal, the appropriate antitoxin may be life-saving.

A toxin with which we all (should) have personal experience is the tetanus toxin produced by *Clostridium tetanii*. Periodically, we are (should be) immunized with formaldehyde-fixed preparation called *tetanus toxoid*. (An inactivated toxin used for purposes of immunization is called a toxoid.) If immunity to the toxin were not established before infection with *C. tetanii*, sufficient amounts of neutralizing antibody could not be generated before death.

Long-lasting immunity to diseases such as poliomyelitis and measles is due to priming of the immune system by vaccination. When an antigen is first encountered, the immune response (the primary response) is slow compared to that occurring after the second and subsequent exposures to antigen (the secondary response). Secondary responses are more effective in controlling infections because they occur more rapidly. We will return to these concepts in Chapter 6.

Constant region functions

Complement fixation

As described in Chapter 4, antibodies bound to a cell surface may initiate the classical pathway of complement activation, lysing the cell. IgM, IgG1, 2, 3, and possibly IgG4, function in this way. Complement is activated when C1q binds simultaneously to the C_H2 domains of two antibodies. Since IgG is monomeric, this requires that two antibodies be bound sufficiently near one another on the cell surface. One molecule of pentameric IgM may bind C1q.

Serum half-life

Immunoglobulin concentrations in body fluids are in a dynamic equilibrium determined by rates of synthesis, degradation, and loss to the environment. The latter is particularly important for IgA, prevalent in mucous and serous secretions. An adult human synthesizes approximately two grams of Ig per day. Macrophages and neutrophils take up Ig via pinocytosis and degrade it. Chemically altered (old) Ig is cleared from the blood more rapidly. Apparently, alterations in structure are signals for removal of Ig from the body. The serum half-lives are different for each of the Ig classes (see Table 5.I).

Immune system Fc receptors

Certain cells possess receptors for the constant region of particular Ig classes. For example, basophils and mast cells have high-affinity receptors binding the C_H4 domain of IgE ($Fc_\varepsilon RI$). This interaction underlies many of the phenomena of allergic reactions (see Chapter 10).

A low-affinity receptor for IgE ($Fc_{\varepsilon}RII$, or CD23) is found on some B cells, activated macrophages and eosinophils. The function of this receptor is unknown.

Three types of receptors for IgG FC have been identified ($Fc_{\gamma}RI$-III, also called CD64, CDw32, and CD16, respectively). All three bind IgG1 and IgG3 best, and only bind weakly to IgG2 and IgG4. $Fc_{\gamma}RI$ binds monomeric IgG, $Fc_{\gamma}RII$ and III bind aggregated IgG. Each receptor is variably distributed among monocytes, macrophages, granulocytes, NK cells, and platelets. Fc receptor occupancy may influence intracellular metabolism and the rate of cell division, and is also the basis of *opsonization*. Fc receptor-binding of an antibody whose combining site is fixed to the surface of a cell (e.g., a bacterium), initiates or enhances the process of phagocytosis. The Greek root of this work, *opsonein*, means "to prepare food" (see Chapter 4). B cells also possess $Fc_{\gamma}RII$. Fc receptor-binding may influence (positively or negatively) B cell activation, proliferation, and differentiation (see Chapter 6).

Placental transfer

During gestation and in the first three months of life, humans synthesize only minute amounts of immunoglobulin. This is due to the immune system's immaturity, and lack of immunogenic stimuli during gestation. The placental barrier against mixing of the maternal and fetal circulations protects the fetus from exposure to many antigens and microbes encountered by the mother, although a number of pathogens are able to cross the placenta (e.g., rubella, toxoplasma). In addition to its functions in nutrition, gas exchange, and waste removal, the placenta also actively transports maternal antibodies into the fetal blood. Only IgG is carried across the placenta. This process requires binding of the IgG Fc, IgG Fab is not transported. Placental transfer of maternal antibodies provides the fetus with a high concentration of antibodies in its blood at a time when it is not equipped to synthesize its own.

After birth, an infant continues to be supplied with maternal antibodies via colostrum and milk. Colostrum is the first milk secreted by the mammary glands following parturition, and is rich in SIgA, IgE and IgM. This provides passive immunity to the infant's gut. Trace amounts of Ig are also absorbed into the circulation. Maternal antibodies transferred to the fetus and neonate are an important source of passive immunity in an individual yet unable to mount a vigorous

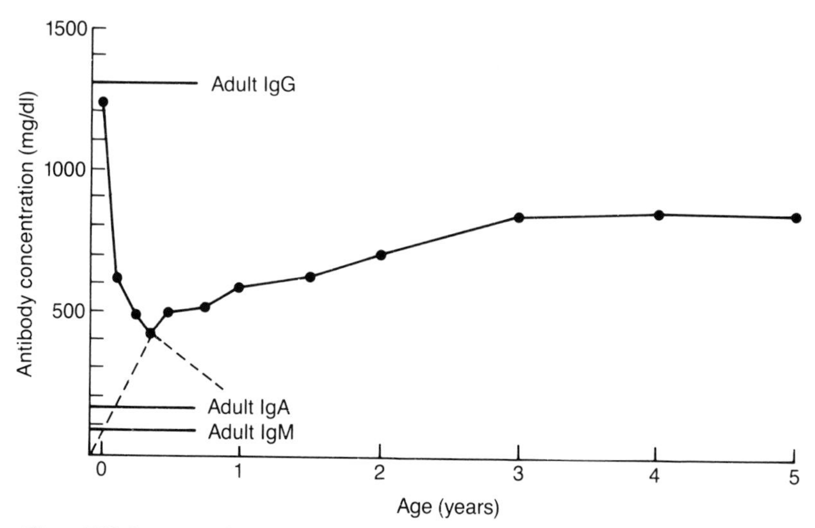

Figure 5.19. *Serum IgG levels in the first five years of life.* Placental transfer of IgG provides the neonate with a serum level that is only slightly below those of the mother. These antibodies are very short-lived, however, decreasing by 50% in two to three months after birth. The infant's synthesis of IgG becomes greater than loss of the mother's by four to six months. Serum IgG levels increase gradually in the first three years, then plateau at about 60% of the adult concentration, while the rate of increase becomes very slow. Normal adult levels are not reached until adolescence. IgA and IgM are not transported across the placenta and the neonate has only minute quantities in its circulation (however, it receives large amounts of IgA in colostrum and mother's milk). Serum levels of IgA and IgM increase gradually and slowly, adult levels also being reached in adolescence. (Adapted from Johansson and Berg, 1967.)

immune response. Figure 5.19 shows the pattern of serum IgG levels in the first five years of life.

Protein A

Some strains of Staphylococci produce a polypeptide known as protein A which binds with varying affinity to the Fc of particular Ig classes of various species. This interaction has been exploited in the purification of Ig for both clinical and research purposes.

MONOCLONAL ANTIBODIES

We close this chapter with a discussion of a technology that has revolutionized research in many biological disciplines, as well as the diagnosis and treatment of disease. What is even more impressive is the youth of this technology, and its tremendous potential which we have only begun to appreciate.

A series of Ig gene rearrangements gives rise to a B cell expressing unique Ig genes. As the B cell divides, all of its progeny cells carry the same unique Ig genes. These genetically identical cells constitute a *clone*, and the antibodies they secrete are all identical, hence, *monoclonal*. When we generate an immune response to a particular epitope, it is *polyclonal*. That is, we have several clones of B cells capable of interacting with that epitope which may be activated during the immune response. We will return to this discussion in Chapter 6.

Myelomas (malignant B cells) of murine origin were an important early source of material for studies of Ig structure. The large amounts of antibodies secreted by myelomas are homogeneous (monoclonal) and easily subject to structural analysis. Myeloma cells may be grown by serial passage in animals, or in some cases, they may be adapted to growth *in vivo*. Occasionally, non-secreting myelomas are isolated, or non-secreting variants of established lines arise after many passages.

In the mid 1970's methods were developed to fuse myeloma cells with normal B cells yielding *hybridomas*. The first hybridomas were products of fusions of antibody-secreting myelomas with normal B cells. These experiments established that the malignant phenotype was preserved in the fused cell, but were not useful for monoclonal antibody production because they synthesized mixtures of antibodies composed of chains from the myeloma and the normal cell. This problem was solved by using a non-secreting myeloma as a *fusion partner* for normal B cells.

A more difficult problem in obtaining monoclonal hybridomas was selecting for growth of fused cells instead of unfused myeloma cells. The answer lay in the phenomenon of *gene complementation*. Suppose that a cell has a genetic defect, and enzyme deficiency due to a mutation in the gene encoding the enzyme. If we introduce into the cell a functional copy of the gene, enzyme production is restored. Thus, if we obtain a myeloma with a selected enzyme deficiency, we can use gene complementation by fusion with a normal B cell to select for growth of fused cells.

Non-secreting myeloma lines were established with a defect in the enzyme *hypoxanthine-guanine phosphoribosyltransferase* (*HGPRT*). This enzyme operates in the salvage pathway of purine nucleotide synthesis, and permits a cell to use xanthine and guanine as nucleotide precursors, rather than synthesizing them *de novo*. A cell with this

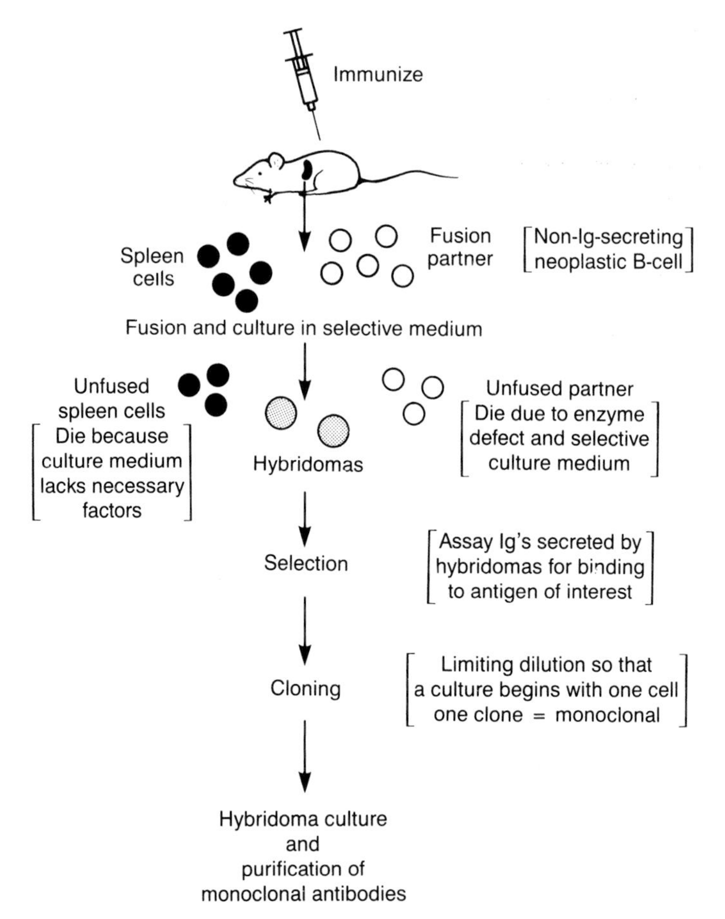

Figure 5.20. *Production of monoclonal antibodies.* An animal is immunized with the antigen of interest. Spleen cells (usually, but lymph nodes are also good sources of lymphocytes) are then removed and fused with the myeloma partner. Unfused myeloma cells do not survive because they are deficient in nucleotide synthesis and cannot survive in the selective medium (see text). Unfused lymphocytes die in culture because the medium lacks necessary trophic and nutrient factors. Hybridoma culture supernatants are commonly screened by solid-phase immunoassays for binding to the antigen of interest. When antigen-binding antibodies are found, the hybridomas are *cloned* by *limiting dilution.* The cells are diluted so that a culture may be initiated with only one cell. This ensures that all of the cells it gives rise to will be genetically identical, one cell = one clone = monoclonal. Monoclonal antibodies may be purified by a variety of techniques (see Figure 5.21).

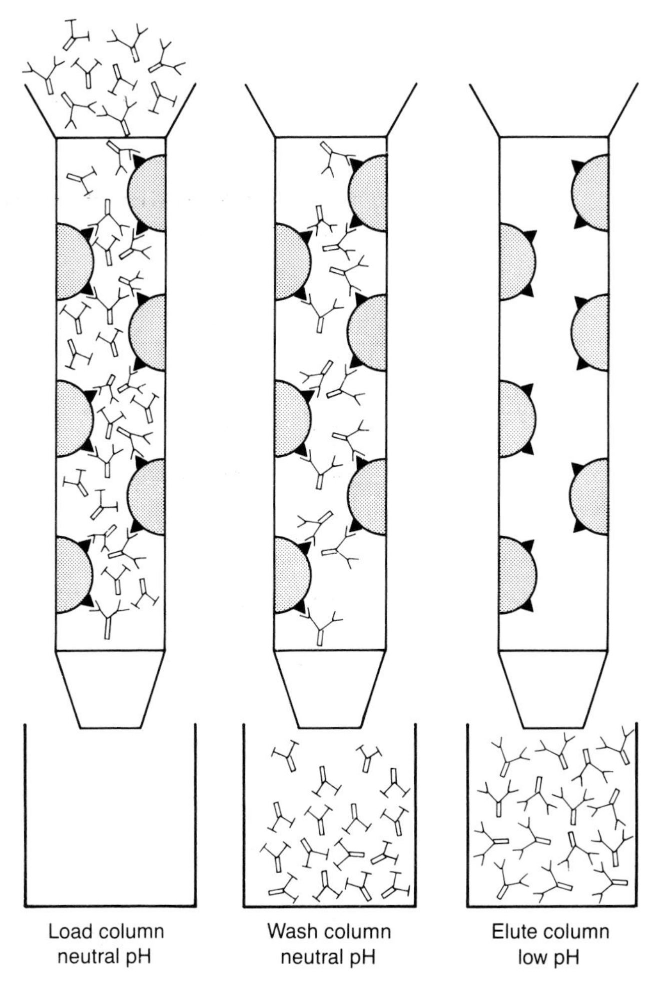

Figure 5.21. *Affinity chromatography.* This technique can be used in a multitude of situations where purification of antibodies, both polyclonal or monoclonal, is desired. The method is applicable to purification of antigens, as well.

For antibody purification, antigen molecules (black triangles) are covalently attached to solid beads which are used to fill a chromatography column. The antibody mixture is loaded at neutral pH (left). Antigen-specific antibodies adhere to the beads, other antibodies pass through. Washing the column at neutral pH removes all non-specific antibodies (center). Washing the column at low pH weakenes the antigen-antibody interaction, and antigen-specific antibodies elute from the column (right).

Clearly, this method works just as well if the desired antibodies are polyclonal or monoclonal. The situation may even be reversed. We may use an antigen-specific antibody to purify a particular antigen from a mixture. We may even use antibodies to purify other antibodies. Isotype, allotype, or idiotype-specific antibodies can be used to select Ig having the desired characteristic.

enzyme defect will not die as a result, but if it is cultured in the presence of an inhibitor of *de novo* nucleotide synthesis (such as aminopterin) it will die since it will be unable to make any purine nucleotides. When a normal cell is fused with a HGPRT-deficient cell, the normal HGPRT gene is expressed. When a mixture of fused and unfused cells are grown in the presence of aminopterin and hypoxanthine, fused cells survive because they use hypoxanthine to make purine nucleotides. Thymidine is added to the culture medium since its synthesis is also inhibited by aminopterin. This selection medium is referred to as *HAT medium* (hypoxanthine, aminopterin, thymidine). Unfused myeloma cells die, and unfused B cells do not grow for extended periods *in vitro* if the culture medium is not supplemented with trophic factors. This scheme is summarized in Figure 5.20. A variety of techniques have been developed for the purification of antibodies. One of the most powerful and popular is *affinity chromatography* (Figure 5.21).

Originally, cell fusion was achieved with syncytia-forming viruses (such as SV40) which possess surface glycoproteins promoting plasma membrane fusion. The most widely applied technique today uses the chemical polyethylene glycol. Another method that has been used successfully is the application of brief pulses of high-voltage electric fields to cell mixtures (electroporation).

The beauty of the hybridoma technology is that one can obtain large quantities of homogeneous (monoclonal) antibody *of any desired specificity*. Let us say one requires antibody against protein X. Immunize a mouse with protein X, remove its spleen cells and fuse them with a myeloma. After HAT selection, use an ELISA or RIA to screen surviving cells for secretion of antibody binding X. We concede this scenario to be somewhat glib, and many pitfalls may be encountered, but we hope that the reader will appreciate the great power of this technology. Following this scheme, the different monoclonal antibodies produced by researchers worldwide number in the hundreds of thousands.

Having examined in some detail the effector molecules of humoral immunity, we turn our attention now to the cells which produce them: B lymphocytes.

SOURCES AND SUGGESTED ADDITIONAL READING

Brandtzaeg, P. (1985) Role of J chain and secretory component in receptor-mediated glandular and hepatic transport of immunoglobulins in man. *Scand. J. Immunol.*, **22**:111–146.

Burton, D. R. (1985) Immunoglobulin G: functional sites. *Mol. Immunol.*, **22**:161–206.

Capra, J. D. & Edmundson, A. (1977) The antibody combining site. *Sci. Am.*, **236**:50–59.

Edmundson, A. B., Ely, K. R., Abola, E. E., Schiffer, M. & Panagiotopoulos, N. (1975) Rotational allomerism and divergent evolution of domains in immunoglobulin light chains. *Biochemistry*, **14**:3953–3961.

Getzoff, E. D., Geysen, H. M., Rodda, S. J., Alexander, H., Tainer, J. A. & Lerner, R. A. (1987) Mechanisms of antibody binding to a protein. *Science*, **235**:1191–1196.

Geysen, H. M., Tainer, J. A., Rodda, S. J., Mason, T. J., Alexander, H., Getzoff, E. D. & Lerner, R. A. (1987) Chemistry of antibody binding to a protein. *Science*, **235**:1184–1190.

Grubb, R. (1970) *The Genetic Markers of Immunoglobulins*, Springer-Verlag, New York.

Hanson, L. A., Ahlstedt, S., Andersson, B., Carlsson, B., Fallstrom., S. P., Mellander, L., Porras, O., Soderstrom, T. & Eden, C. S. (1986) Protective factors in milk and the development of the immune system. *Pediatrics*, **75**:172–176.

Honjo, T., Alt, F. W. & Rabbits, T. H., eds. (1989) *Immunoglobulin Genes*, Academic Press, London.

Horini, T. & Terasaki, P. (1982) Autoantiidiotypic antibody against DR antibody. *Hum. Immunol.*, **5**:144–149.

Ichihara, Y., Matsuoka, H. & Kurosawa, Y. (1988) Organization of human immunoglobulin heavy chain diversity gene loci. *EMBO J.*, **7**:4141–4150.

Johansson, S. G. & Berg, T. (1967) Immunoglobulin levels in healthy children. *Acta Pediatr. Scand.* **56**:144–149.

Kabat, E. A., Wu, T. T., Reid-Miller, M., Perry, H. M. & Gottesman, K. S. (1987) *Sequences of Proteins of Immunological Interest.* Fourth edition, U. S. Department of Health and Human Services, Public Health Service, National Institutes of Health.

Köhler, G. (1986) Derivation and diversification of monoclonal antibodies. *Science*, **233**:1281–1286.

Natvig, J. B. & Kunkel, H. G. (1973) Human immunoglobulins: classes, subclasses, genetic variants, and idiotypes. *Adv. Immunol.*, **16**:1–59.

Plotkin, S. A. & Mortimer, E. A., Jr., eds (1988) *Vaccines*, W. B. Saunders Company, Philadelphia.

Porter, R. R. (1959) The hydrolysis of rabbit gamma-globulin and antibodies with crystalline papain. *Biochem. J.*, **73**:119–126.

Potter, M. (1967) The plasma cell tumors and myeloma proteins of mice. *Methods Cancer Res.*, **2**:106–131.

Rose, N. R., Friedman, H. & Fahey, J. L., eds. (1986) *Manual of Clinical Laboratory Immunology*, American Society for Microbiology, Washington, D. C.

Tiselius, A. & Kabat, E. A. (1939) An electrophoretic study of immune sera and purified antibody preparations. *J. Exp. Med.*, **69**:119–131.

Waldmann, T. A., Strober, W. & Blaese, M. (1970) Variations in the metabolism of immunoglobulins measured by turnover rates. In *Immunoglobulins*, E. Merler, ed., National Academy of Sciences, Washington, D. C.

Wu, T. T. & Kabat, E. A. (1970) An analysis of the sequences of the variable regions of Bence Jones proteins and myeloma light chains and their implications for antibody complementarity. *J. Exp. Med.*, **132**:211–250.

Yalow, R. S. & Berson, S. A. (1960) Immunoassay of endogenous plasma insulin in man. *J. Clin. Invest.*, **39**:1157–1175.

Chapter 6

B Cells and Humoral Immunity

B CELL DEVELOPMENT

B lymphocyte differentiation is divided into several stages: *stem cell*; *progenitor cell* or *null cell*; *pre-B cell*; *immature B cell*; *virgin B cell* or *mature B cell*; *B lymphoblast*; *plasma cell*; and *memory cell*. An outline of B cell differentiation is shown in Figure 6.1.

As described in Chapter 2, stem cells give rise to all cellular elements of blood. A complex interplay of intracellular and extracellular regulatory mechanisms determines when a stem cell enters a particular differentiative pathway. We may operationally define a B cell developmental stage where the stem cell has "committed" itself to become a B cell, and thereby loses the capacity to become any other type of cell. This stage has been called *B progenitor cell*, *null B cell*, or *pre-pro-B cell*. The earliest "B cell-specific" characteristic identified is rearrangement of immunoglobulin heavy chain D and J_H genes (see Chapter 5). D-J rearrangement in at least one chromosome is a marker of the null B cell.

After D-J rearrangement, a V_H gene is joined to the newly formed DJ construct. Following this event, complete μ heavy chain mRNA is transcribed and translated. The IgM heavy chain remains in the cytoplasm and is not secreted or integrated into the cell membrane. Cytoplasmic μ chains mark the *pre-B cell* stage. At this point, B cells first acquire receptors for various B cell mitogens (stimulators of mitosis) such as bacterial endotoxins (see below).

Following successful assembly of an Ig light chain and its combination with the heavy chain, a complete IgM antibody molecule appears on the cell surface. This signals progression to the *immature B cell* stage. In addition to surface IgM, immature B cells also express receptors for Fc of IgG (see Chapter 5), and class I and class II histocompatibility antigens (see Chapter 8).

Appearance of IgD (in addition to IgM) on the cell surface marks arrival at the *mature* or *virgin B cell* stage. Cells at this stage also express

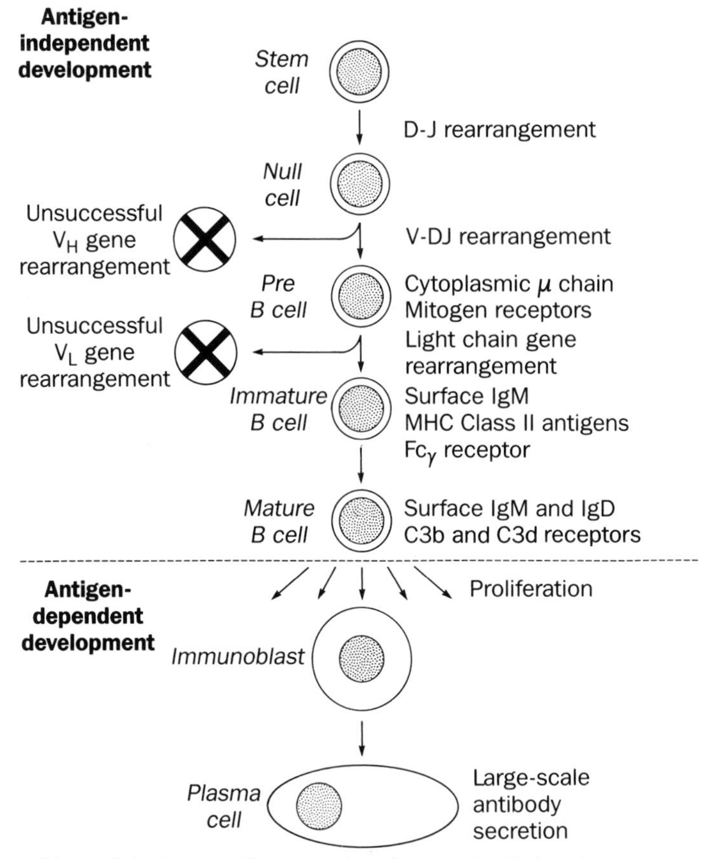

Figure 6.1. *B cell development.* The generation of mature B cells from bone marrow stem cells is independent of antigen stimulation. Various stages of development are marked by rearrangement and expression of immunoglobulin genes, and presence of particular surface glycoprotein antigens and receptors. Further development of mature B cells depends on antigen stimulation. Instead of becoming a plasma cell, a B cell may become a memory cell.

receptors for the C3b and C3d components of complement (see Chapter 4). Up to this point in B cell differentiation, progress does not require stimulation by antigen. Thus, these stages are often called the *antigen-independent phase* of B cell development. Mature B cells which have not been activated are also called *resting B cells.*

This is a good point at which to return to consideration of immunoglobulin genetics. The reader may have speculated that since there are two copies of each chromosome within a cell, a B cell might produce two different heavy chains or two different light chains. This could result in formation of several different antibodies in a single cell. In fact, this does *not* occur.

During the null B cell stage, D-J rearrangement usually occurs in both heavy chain loci before V-DJ joining. If the first V-DJ rearrangement is successful, a complete heavy chain mRNA is made, and cytoplasmic μ chain begins to accumulate. In addition to triggering rearrangement of light chain genes, the μ protein may be a negative signal inhibiting further heavy chain gene rearrangement on the other allele. If the first heavy chain rearrangement is unsuccessful, the other locus will rearrange. If both rearrangements are unsuccessful, no antibody can be formed.

κ light chain loci rearrange before λ loci. Successful κ gene rearrangement allows formation of a complete antibody which seems to shut down further V gene rearrangement. If κ rearrangement is unsuccessful (in both alleles), λ genes then recombine. If no rearrangement produces a functional light chain, again, no antibody can be formed. Thus, a single B cell will produce only a single antibody specificity resulting from the combination of one heavy chain with one light chain. These complex genetic mechanisms underlie a phenomenon long known in immunology, that of *allelic exclusion.* In an individual heterozygous with respect to a particular heavy or light chain isotype, a given B cell will produce antibodies of only one allotype.

It has been mentioned that the co-expression of IgM and IgD on the cell surface markes the mature B cell stage. Recalling the organization of immunoglobulin heavy chain genes (Figures 5.3 and 6.5), this may seem obscure since the C_μ gene intervenes between VDJ and the C_δ gene. Simultaneous expression of a single VDJ associated with C_μ or C_δ occurs by differential splicing of mRNA (Figure 6.2).

Following the antigen-independent phase of B cell development, we enter the *antigen-dependent phase.* These are developmental stages following encounter of the B cell with stimuli leading to antibody secretion. The first visible event in activation is *blast transformation.* The B cell enlarges and begins to divide. While resting mature B lymphocytes are relatively small, only 7–9 μ in diameter, the rapidly dividing *B lymphoblast* or *immunoblast* is much larger, about 15 μ in diameter. Following this brief phase of rapid cell division, the B cell makes a decision: either it will become an antibody-secreting cell or *plasma cell*, or it will become a *memory cell.*

The plasma cell is large, up to 20 μ in diameter, and secrets antibodies prodigiously, but has few on its surface. Its characteristic appearance

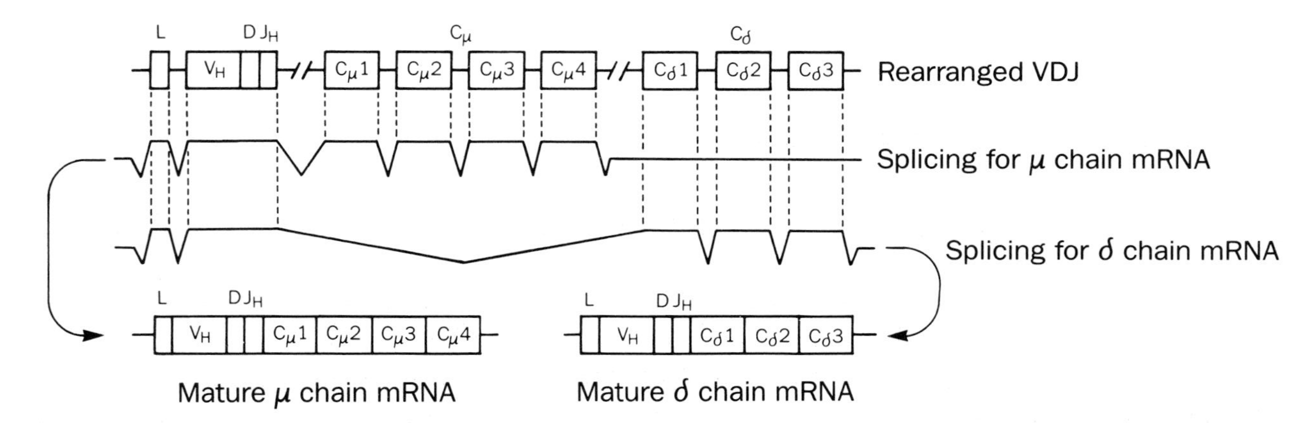

Figure 6.2. *Differential splicing of μ and δ mRNA.* A single primary RNA transcript extending over both μ and δ loci, if spliced in the two different ways shown, may generate a mature mRNA encoding the IgM heavy chain, or a different one encoding IgD. The IgD hinge region is encoded by a separate exon between $C_\delta 1$ and $C_\delta 2$ (not shown).

makes it relatively easy to identify in micrographs (Figures 6.3 and 6.4).

The memory B cell is an elusive creature which is morphologically indistinguishable from other types of resting B cells. It is not yet clear what factors influence a B cell to become a memory cell rather than a plasma cell. Memory cells are more sensitive to antigen stimulation, and proliferate and generate plasma cells more rapidly than do mature B cells. Memory cells are responsible for the more rapid production of antibodies in the secondary immune response (see below).

At some point after the mature B cell stage is reached, an even called *class-switching* may occur. From production of IgM and IgD, a cell (and, thenceforth, all of its progeny) may switch to the production of IgG, IgE, or IgA. Class-switching is influenced by T cell-derived cytokines, but it may occur to some extent in their absence.

What genetic mechanisms underlie the class-swtich? We have already seen how a single cell can produce both IgM and IgD by differential splicing of mRNA (Figure 6.2). The most widely held view of the mechanism of class-switching is somewhat different, and is very similar to the process of V-D-J rearrangement. In the DNA preceding (5′ to) the various constant region genes (except δ) are areas called *switch regions* or *switch sequences.* Class-switching occurs by joining the μ switch region between J_H and C_μ to any of the other switch sequences. DNA between the two switch regions is excised.

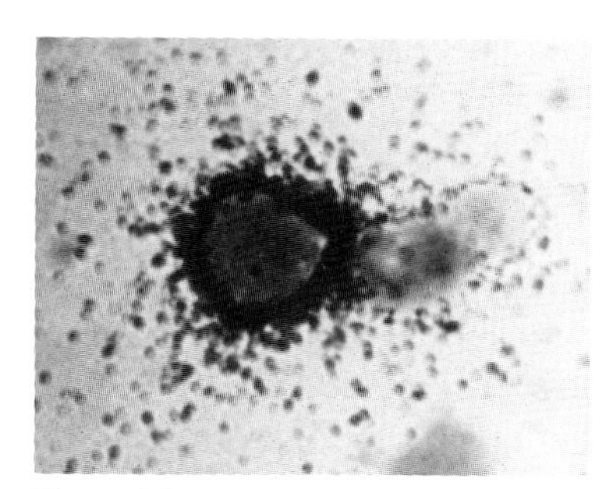

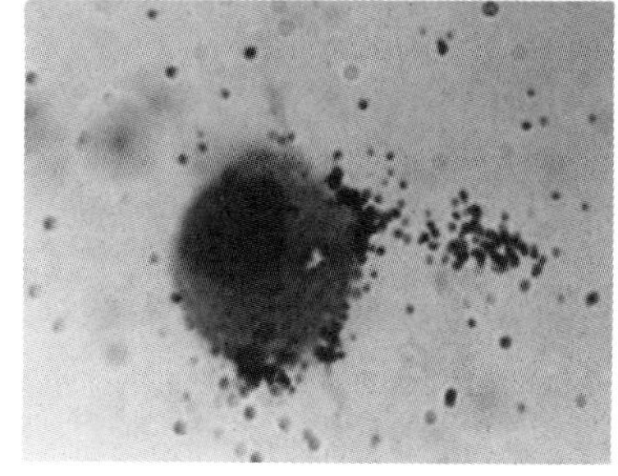

Figure 6.3. *Autoradiography of antibody-bearing cells.* Mice were immunized with *S. enteritidis* lipopolysaccharide. Spleen cells were incubated *in vitro* with LPS biosynthetically labelled with ^{14}C. These cells were then fixed on a slide and stained. The slide was subsequently coated with a photographic emulsion. Radioactive emissions of the LPS caused deposition of silver grains.

A. *B lymphocyte.* This cell is ringed with dense silver grain deposits indicating the large amount of LPS-specific immunoglobulin on its surface.

B. *Plasma cell.* A typical large plasma cell with a lightly-staining round nucleus at the periphery, and abundant cytoplasm. The silver grain density is much lower about this cell, indicating a paucity of Ig receptors relative to the small lymphocyte.

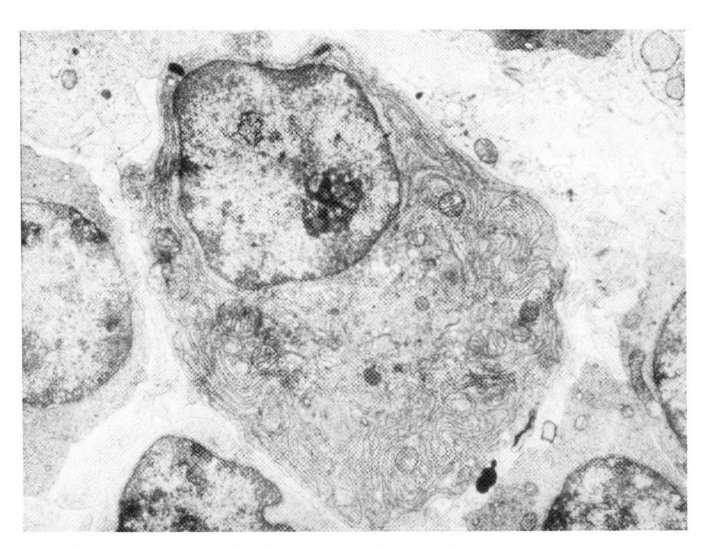

Figure 6.4. *Electron micrograph of a plasma cell.* In addition to several mitochondria, the cytoplasm contains abundant rough endoplasmic reticulum with cisternae distended by an amorphous homogeneous material (immunoglobulin).

Two additional mechanisms may also account for class-switching, and the extent to which they contribute to this phenomenon has not been resolved. One hypothesis is identical to the differential splicing of a transcript containing μ and δ. If we imagine that RNA polymerase may continue along the DNA transcribing all of the C region genes, then VDJ joined to any isotype can be generated by differential splicing.

Another model for class-switching invokes *unequal sister-chromatid exchange* (Figure 6.5). During mitosis, sister chromatids may exchange some portions of themselves. If the positions of the joints between chromatids is identical, one has *homologous recombination*, or *equal crossing over*. The amount of genetic information in each chromatid remains unchanged. If the breaks occur in different positions in the chromatids, one has *non-homologous recombination*, or *unequal crossing over*. This results in deletion of genetic information from one chromatid, and its duplication in the other.

Since class-switching yields the same VDJ in association with a different C region, only isotypic and allotypic determinants are changed relative to the antibody that was produced before the switch. Antigen specificity and idiotypic determinants remain unchanged after class-switching.

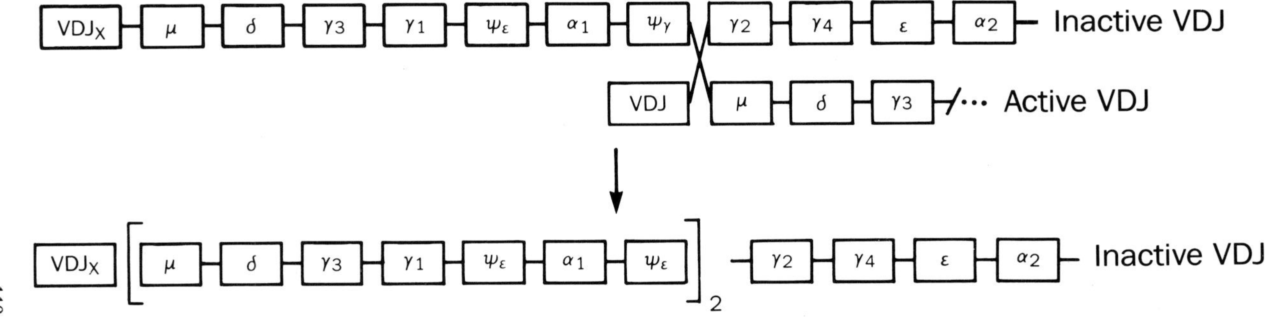

Figure 6.5. *Immunoglobulin class-switching via unequal sister-chromatid exchange.* On one chromosome (inactive), an aberrant rearrangement cannot be expressed, on the other (active), a productive rearrangement is expressed as IgM and IgD. During mitosis, a recombination takes place between the μ switch region of the active chromosome, and the $\gamma 2$ switch region of the inactive allele. This duplicates the sequences in brackets on the inactive chromosome, with concomitant deletion from the active allele. The productive VDJ rearrangement is now 5′ to the $C_{\gamma 2}$ gene. Note that this C gene (and the other C genes 3′ to it) were initially on the inactive chromosome. The Greek letter ψ (psi) designates pseudogenes.

Our final look at immunoglobulin genetics concerns the mechanism by which a cell makes both membrane and secreted forms of an antibody. Cell-surface antibodies are anchored in the membrane by a hydrophobic peptide. Secreted immunoglobulins have a different carboxyl terminus. If you ventured a guess that differential mRNA splicing is involved, you are correct. C region genes have coding sequences for the carboxyl termini of both membrane and secreted forms of immunoglobulins. Messenger RNA for either form can be generated by different splicing of the same primary RNA transcript (Figure 6.6).

Figure 6.1 summarizes the origin of one B cell *clone*. That is, the diagram shows the possible outcomes of a single B cell's attempt to productively rearrange Ig genes and synthesize antibody. Unsuccessful cells are lost from view (they die). Successful cells become part of the circulating pool of resting B cells, and are available to participate in immune responses. All progeny B cells generated by mitosis of a cell having undergone one particular series of rearrangements is called a *clone*, since they are genetically identical. Some genetic divergence may occur since an individual cell of a clone may class-switch or mutate its V genes differently from its siblings. Depending on one's point of view, one may designate the progeny of these genetic events new clones, yet they all remain *clonally related* since they originated from one cell. Thus, each series of immunoglobulin gene rearrangements defines a

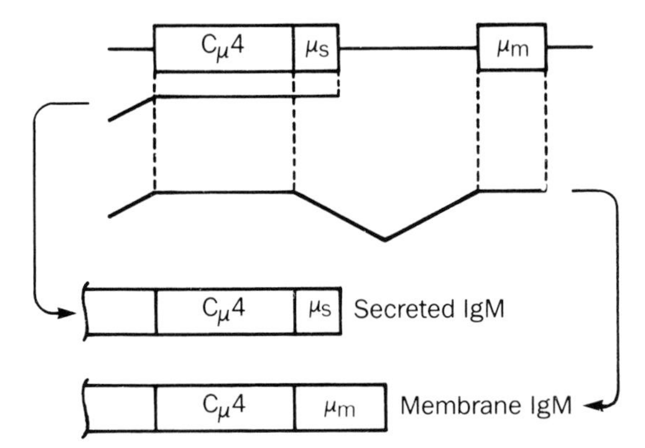

Figure 6.6. *Differential splicing of mRNAs encoding membrane and secreted forms of IgM.* The carboxyl terminus of secreted IgM (μ_s) is encoded by the $C_\mu4$ exon. The carboxyl terminus of membrane IgM (μ_m) is encoded by a separate exon a short distance 3′ to $C_\mu4$. Differential splicing of the same primary transcript generates mRNA encoding either μ_s or μ_m.

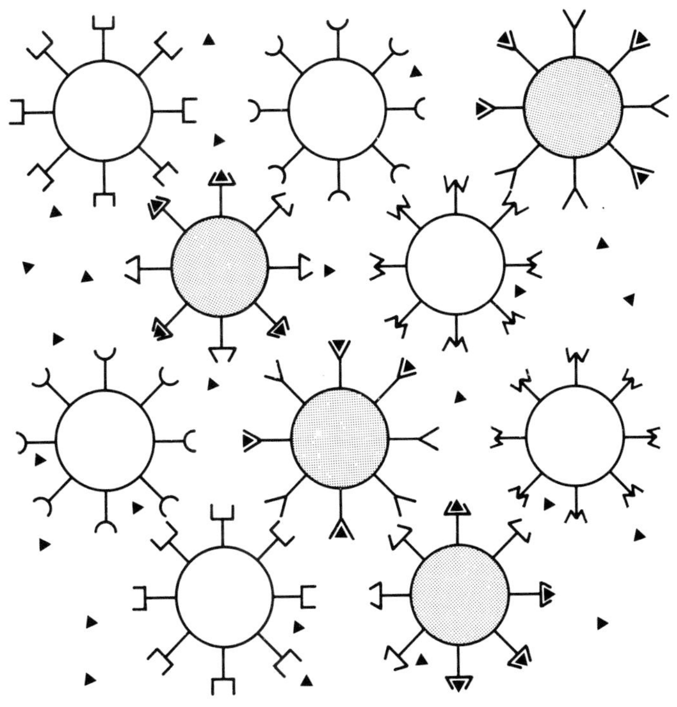

Figure 6.7. *Clonality of circulating B cells.* This group of B cells contains five distinct clones, that is, five different groups of B cells, all members of a group being genetically identical and having the same antigen specificity. Each population derives from a single progenitor cell which underwent a unique series of immunoglobulin gene recombinations. When an antigen (black triangles) enters the circulation, it interacts with all clones capable of recognizing its epitopes. In this example, two of the five clones (the shaded cells) possess receptors which recognize antigen determinants. These cells will produce the antibodies which constitute the humoral response. Hence, the response is *polyclonal*, derived from more than one clone.

new clone, and the circulating pool of B lymphocytes is comprised of many genetically distinct clones, each of which may have anywhere from one to thousands of members. Figure 6.7 further illustrates this concept.

Characteristics of mature B cells

As explained in Chapter 2, it is impossible to distinguish B and T lymphocytes by purely morphological criteria. However, as we have alluded to previously, B cells express several surface molecules, some unique to B cells, some not. Receptors for Ig Fc and complement components, or major histocompatibility (MHC, see Chapter 8) class II antigens are expressed by B cells and other cell types (such as macrophages). Three classes of molecules have been identified which are unique to mature B cells: immunoglobulins, certain cytodifferentiation antigens, and receptors for B cell mitogens. Having dealth with immunoglobulins at some length, we will now focus briefly on the latter two types of B cell markers.

B cell CD antigens

Several cytodifferentiation antigens have been identified on human B cells. The B cell marker expressed earliest in differentiation (before cytoplasmic μ chains) is CD 19, its function is unknown. CD20 is present on $>95\%$ of B cells in both peripheral blood and lymphoid organs. It has been suggested that this molecule may function as a membrane ion channel important in B cell activation. CD21 (also called CR2) is the receptor for the C3d complement fragment and Epstein-Barr virus (EBV). CD21 is found on the majority of lymphoid organ B cells, but is infrequent among peripheral blood B cells. B cells also express CD23 which may function as a receptor for IgE Fc. Finally, B cell CD40 may be a receptor for an as yet unidentified growth factor.

B cell mitogen receptors

Some molecules of microbial or vegetal origin bind to B cell receptors and stimulate their proliferation and differentiation. These substances are *polyclonal activators* since they stimulate B cells irrespective of their antigen specificity. Not all B cells have a receptor for a particular mitogen.

In general, B cell mitogens are polymers having a repeating structural unit. Many are polysaccharides of bacterial or plant origin, or plant lectins, proteins which agglutinate cells or precipitate certain macromolecules. By virtue of their repeating units, these substances presumably cross-link cell-surface receptors and deliver a stimulatory signal. Mitogenic effect is commonly assessed by examining the rate of DNA synthesis. A radioactive substance such as tritiated thymidine, when present in the culture medium, is incorporated into DNA, and can be

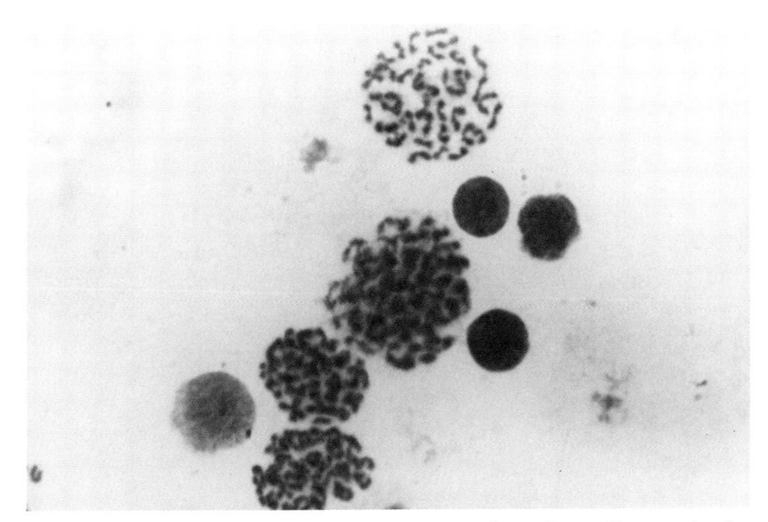

Figure 6.8. *Mitogen-induced B cell proliferation.* Murine spleen cells were incubated *in vitro* with *E. coli* lipopolysaccharide. The cells were then fixed and subjected to the Feulgen reaction which stains DNA. Several cells in this micrograph are clearly in mitosis (metaphase).

measured in a radiation counter. Mitogenic effect may also be assessed visually. Mitogen-treated cells can be fixed and stained, and the number of actively dividing cells counted in a microscope (Figure 6.8).

Numerous B cell mitogens have been identified. Some examples are bacterial lipopolysaccharide or endotoxin, dextran (glucose polymer), levan (fructose polymer), galactan (galactose polymer), pneumococcal capsular polysaccharide, and polymerized flagellin (a protein found in bacterial flagellae). Mitogens may be effective for cells of only one or a few species. Two potent mitogens for human B cells are a lectin isolated from pokeweed (*Phytolacca americana*) hence, *pokeweed mitogen* (*PWM*), and *Nocardia water-soluble mitogen* (*NWSM*) isolated from actinomycetes of the genus *Nocardia.*

B cell subsets

Although all mature B cells have immunoglobulin receptors and certain other cell surface markers in common, they may be divided into subgroups based on expression of particular cytodifferentiation antigens. Furthermore, these subsets are functionally distinct, and may have different patterns of activation in immune responses.

One recently described distinct subset of B cells bear a cytodifferentiation antigen previously considered to occur only on T cells, CD5. These cells have characteristics of immature B cells (IgM^+, IgD^-), have low surface density of MHC class II antigens, and are found in highest frequency in immature stages of development (in the blood of the fetus and neonate). This subset apparently contains precursors of cells producing autoantibodies (antibodies reacting with self antigens) which may cause disease. This subset is found with abnormally high frequency in mouse strains such as NZB which develop spontaneous autoimmune diseases. In humans, it has been shown that rheumatoid factors and anti-DNA autoantibodies are exclusively produced by $CD5^+$ B cells. The special role (if any) of $CD5^+$ B cells in the immune system is not known.

HUMORAL IMMUNITY

Let us begin by distinguishing *passive immunity* and *active immunity*. Passive immunity is the acquisition of preformed antibodies from an external source. For example, the most common infection transmitted by accidental needle-stick injuries in the hospital is hepatitis B. Anyone experiencing such an accident in a high-risk situation (e.g., an intravenous drug-abusing patient) should immediately be given an injection of *hepatitis B immune globulin.* These antibodies will bind to the hepatitis B virus and prevent it from initiating an infection. Our own immune systems would be incapable of generating a sufficiently large antibody response quickly enough to achieve this.

Passive immunity is not only iatrogenic, but is an important physiological mechanism protecting neonates. As described in Chapter 5, the passive immunity we acquire via placental transfer of IgG and via colostrum is invaluable in protecting infants from infection.

As one might guess, active immunity is the response generated during encounter of the immune system with antigen. This may occur during the course of a natural infection, or after intentional antigen administration (vaccination).

B cell activation by antigen

Vis-a-vis the antibody response, antigens may be categorized as *thymus-independent type 1*, *thymus-independent type 2*, or *thymus-*

dependent (*TD*) antigens (also see Chapter 3). The first two groups are often referred to simply as *type 1* or *type 2* antigens. Although these names serve as convenient descriptive labels when discussing mechanisms of B cell activation, they should not be considered restrictive or mutually exclusive. An antigen may behave as type 1, type 2, or TD depending on its chemical state or associated molecules. An antigen may also exhibit properties of more than one class depending on the dose administered.

In discussions of the antigen-dependent phases of B cell development, several distinct processes may be distinguished. The first is *activation*. In this context, activation specifically denotes the resumption of the cell cycle, or the exit of the cell from its resting state. The consequence of activation is *proliferation*, since the cells are now actively dividing. Proliferation is simply an increase in cell number. *Differentiation* denotes the changes occurring in B cells that enable large-scale antibody production. It is the transformation from the resting B cell phenotype to the plasma cell phenotype. In some instances, trophic factors and other exogenous signals have been shown to specifically influence one of these processes without affecting the others. The generation of many antibody-secreting cells from one resting B cell is called *clonal expansion*.

Type 1 antigens

Substances acting as polyclonal B cell mitogens (described above) comprise the type 1 antigens. Some preferentially activate B cells, some T cells, and some induce proliferation and differentiation of both. At low concentrations, B cell mitogens stimulate specific antibody responses (i.e., only antibodies which bind the mitogen are produced). At higher concentrations, mitogens induce B cell proliferation irrespective of their antigen specificity.

The action of B cell mitogens appears to require some participation of T cells. The mitogen may induce early stages in B cell activation, rendering them receptive to additional stimulatory signals provided by T cells. An exception is Epstein-Barr virus (EBV). This virus is able to transform some B cells (render them malignant). This transformation enables B cells to become antibody-secreting cells independently of the usual regulatory factors. The EBV receptor is also the receptor for the C3d complement fragment (CR2). This receptor is also designated CD21.

The fact that mitogens activate murine B cells without regard to their antigen specificity suggests that these cells bear a "mitogen receptor" distinct from immunoglobulin. Some have speculated that cross-linking of the mitogen receptor by repeating determinants of the mitogen delivers the activating signal. Others have suggested that cross-linking of surface immunoglobulin is the key event in B cell activation, and that cross-linking of mitogen receptors adventitiously cross-links B cell surface Ig. Putative mitogen receptors remain to be characterized, and neither of these models has been unequivocally established. Human B cells may or may not have distinct mitogen receptors (Figure 6.9).

Type 2 antigens

Type 2 antigens differ from type 1 in that the former induce predominantly antigen-specific antibody responses. Type 2 antigens are also molecules with repeating structures, examples are polysaccharides and synthetic polynucleotides. These antigens are able to cross-link specific immunoglobulin receptors on B cells. Surface Ig cross-linking alone is not

▶

Figure 6.9. *B cell activation by type 1, type 2 and TD antigens.*

A. *Type 1 antigens.* These antigens stimulate antibody production independently of specificity. In mice it is apparent that B cells possess (a) distinct receptor(s) for mitogens since several have been shown to act independently of accessory cells or T cells. Whether or not human B cells have such (a) receptor(s) has not yet been determined. Cross-linking of such receptors (with or without simultaneous cross-linking of surface Ig) has been hypothesized by many to be an intermediate step in activation, but has not been clearly demonstrated. For simplicity, we diagram the mitogen interacting with a unique receptor. Depending on the mitogen, subsequent action of cytokines may or may not be required.

B. *Type 2 antigens.* These antigens stimulate specific antibody responses. A stimulating signal is delivered to the B cell by cross-linking surface Ig via the multiple identical epitopes contained in the antigen. Subsequent B cell differentiation depends on the action of cytokines.

C. *TD antigens.* These antigens require processing and presentation to a T_h cell in order to generate specific antibody responses. Processing may be by an antigen presenting cell or a specific B cell. The T_h cell-B cell interaction (cognate help) most probably requires association of antigen with B cell surface MHC class II molecules. A role for the Ig receptor has long been postulated in this interaction, but its importance here is now debated. Since its participation has not been definitely ruled out, it is included here with a question mark indicating uncertainty as to the precise interrelationships of the various participants in the interaction. As with antigen presentation, adhesion molecules are also active in stabilization and/or signal transduction in the cellular contact.

sufficient to induce B cell proliferation and differentiation, but renders B cells receptive to T cell factors which further stimulate their progression toward antibody-secreting cells (Figure 6.9). At very high concentrations, many type 2 antigens begin to behave as type 1 antigens, suggesting a weak interaction with (putative) mitogen receptors.

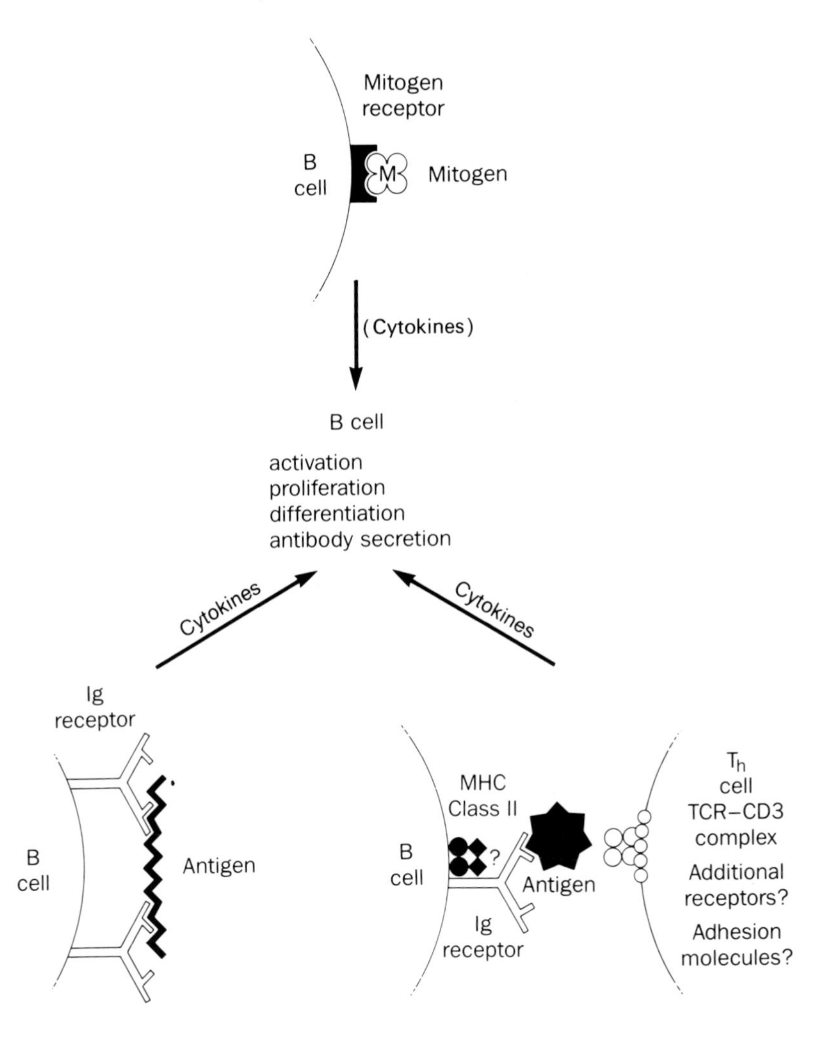

TD antigens

In antibody responses to TD antigens, two types of cellular interaction are required. In the first, antigen is *presented* to a helper T cell. T cells only recognize antigen when it is associated with MHC molecules. This association is called *antigen presentation* and will be described in Chapter 7. The second interaction in B cell responses to TD antigens is between the T_h cell and an antigen-specific B cell.

In generating hapten-specific antibodies after immunization with hapten-carrier conjugates, T_h cells are specific for carrier epitopes, while B cells bind hapten. One long-accepted model of T_h-B interaction held that antigen formed a "bridge" between the B cell Ig receptor and the T cell antigen receptor. The B cell immunoglobulin bound to a *haptenic* or *B cell determinant*, while the T cell receptor bound to a *carrier* or *T cell determinant*. B cell surface MHC class II molecules are also involved in the interaction.

Subsequent to observations that a) T cells recognize antigen-derived peptides associated with MHC molecules; and b) B cells are able to present antigen, the role of immunoglobulin in the T-B interaction has been called into question. While surface Ig may be necessary for efficient antigen presentation by B cells, its requirement in T-B collaboration in TD immune responses has not been established.

Thus, the initial stimulus for B cell activation in TD antibody responses may be a two-contact or a one-contact process. The T_h cell may be activated by an antigen-presenting cell, then interact with a B cell (two contacts). Alternatively, the B cell may present antigen, activate the T cell, and in turn, be activated itself (one contact, Figure 6.9). *Adhesion molecules* (see Chapter 7) are also important in stabilizing the interaction of and/or transducing a signal between T_h and B cells. The T cell-B cell interaction required for antibody responses to TD antigens is often called *cognate T cell help.*

Subsequent steps in proliferation and differentiation of B cells to antibody-secreting cells require several factors produced by T cells and other cell types. These are the *cytokines* (also described in Chapter 7).

B cell activation by immunoglobulins

Anti-immunoglobulin antibodies

Antibodies binding C_μ or C_δ determinants of B cell surface immunoglobulin (anti-IgM or anti-IgD antibodies) may stimulate early steps in B cell activation. Differentiation to antibody-secreting cells, however, requires T cell-derived factors. As might be expected, anti-Ig antibodies activate B cells regardless of their antigen specificity. This activation requires cross-linking of surface Ig, since anti-Ig Fab fragments are not activating, while $F(ab)'_2$ fragments are. Indeed, early experiments demonstrating B cell activation by surface Ig cross-linking led many to believe that such cross-linking is an essential step in activation by antigen. Whether or not surface Ig cross-linking is a "final common pathway" in B cell activation is still a matter of debate. We do not yet known all of the intracellular effects of occupancy or cross-linking of the multitude of B cell surface receptors, including immunoglobulin. As discussed above, Ig cross-linking may be only one type of signal sufficient to render B cells receptive to additional stimuli leading to antibody secretion.

Immune complexes and Fc regions

The literature on this subject is voluminous and complex. Depending on the character of immune complexes (antigen/antibody ratio), the timing of administration (before, with, or after antigen), and the isotype of the antibodies in the complex (IgG versus IgM), immune complexes and/or Fc regions have been shown to either activate or suppress B cells independently of antigen, or stimulate or suppress specific antigen responses.

Antigen-specific antibodies administered passively with antigen may suppress immune responses by masking epitopes and interfering with antigen recognition. Some have also observed that exogenous antibody Fc regions interfere with T-B cell cooperation. In some experiments, however, passively administered antibodies enhance antigen-specific responses. The cross-linking of B cell surface immunoglobulin with B cell Fc receptors (FcR) has been observed to suppress B cell activation. Some have implicated T cell-derived suppressor factors in immune complex-mediated B cell suppression.

Needless to say, we require a much greater knowledge of the interrelationships between B cell Ig and Fc receptors and the consequences of their occupancy and cross-linking one with another before we can make sense of much of the phenomenology of immune complex- and Fc-mediated activation or suppression of antibody responses. In many instances, accessory cells and T cells appear to play a role, as well as complement derived proteins (don't forget, B cells also possess receptors for these molecules).

Cellular events in B cell activation

We have seen that there are several pathways by which resting B cells may be induced to begin cycling. B cells may be activated by binding of type 1 antigens (mitogens) to mitogen receptors, by binding of type 2 antigens to Ig receptors and/or to mitogen receptors, and by appropriate MHC-determined contacts with activated T_h cells. In addition, B cells may be activated by binding of the F_c portion of IgG to the F_c receptor, by binding of aggregated C3d to its receptor (CD21), or by cross-linking surface Ig with anti-Ig antibodies. The precise manner in which all of these different situations activate B cells is not yet known. The cross-linking of surface immunoglobulin either with itself or with another molecule appears to be important in at least some of the above phenomena.

Once an activating signal is delivered, the B cell enters on a differentiative pathway leading either towards a plasma cell, or a memory cell (see below). The pathway consists of a programmed response to a variety of cytokines (see Chapter 7). The interleukins 2, 4, 5, and 6, and interferon-γ have been implicated in the control of various steps along this pathway. Interleukins also influence the relative proportions of various Ig isotypes produced in antibody responses. We do not yet know whether all of these cytokines are required in all mechanisms of B cell activation (probably not) or what the precise sequence of action of the various cytokines may be. Many of the cytokines (and their receptors) have been cloned. Pure molecules and reagents interacting with them produced by recombinant techniques will hopefully allow us to map in greater detail the different steps in B cell activation.

At the intracellular level, surface Ig cross-linking leads first to cleavage of phosphatidylinositol bisphosphate yielding inositoltriphosphate and diacylglycerol (DG). Accumulation of DG is essential for the action of protein kinase C. These events lead to mobilization of intracellular calcium and influx of extracellular calcium. Entry of the cell into G_1 phase is associated with phosphorylation of cytoskeletal proteins.

Antibodies binding MHC antigens deliver an activating signal via cyclic AMP (cAMP). Increased cAMP activates protein kinase A, which then induces relocalization of intracellular protein kinase C. Activation of only one of these intracellular pathways appears insufficient to initiate cell cycling, both are required.

Additional biochemical pathways may operate in B cell activation. Whether the two described above, along or together, or in combination with other pathways, are operating in a given circumstance probably depends on the specific activating mechanism (antigen, immune complex, anti-Ig antibody, mitogen, etc.).

Immunologic memory

The immune response to a TD antigen in an animal which has never before encountered the antigen is called a *primary immune response.* The response of an animal which has been *primed* by at least one prior exposure to antigen is called a *secondary immune response.* Primary and secondary responses differ with respect to the *lag period* (the period of time between exposure to an antigen and an observable immune response), the magnitude of the response, and the isotypes and affinity of antibodies produced (Figure 6.10).

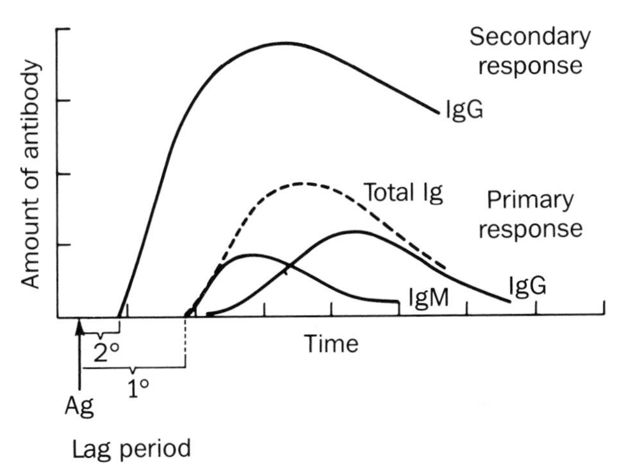

Figure 6.10. *Primary and secondary antibody responses to TD antigens.* Antigen (Ag) is administered at the time indicated. The lag period of the primary (1°) response is longer than that in a secondary (2°) response. In the 1° response IgM is initially produced, followed later by IgG. The total antibody produced is represented by the dotted line. The time scale is not marked since the course of the response will vary greatly depending on the dose of antigen, the route of administration, and how rapidly it is cleared from the body. For many protein or glycoprotein antigens not associated with an ongoing infectious process, the time course of the response ranges from two to four weeks. In the 2° response, the initial rise in antibody production is much more rapid and the total amount produced greater than in the 1° response. In addition, there is a predominance of the IgG isotype (although IgM antibodies are produced as well), and the antibodies have higher affinity for antigen.

The antigen concentration is important in initiating an immune response. There is a threshold level below which no response is obtained. Above that threshold, the response increases with increasing doses, but less than proportionately. Excessive antigen doses may lead to tolerance (see Chapter 9). Following antigen administration, there occurs the lag period before antibody is detected. This period corresponds to the cellular events in the activation, proliferation, and differentiation of B cells to antibody-secreting cells.

The *kinetics* of the antibody response (the changes in antibody production with respect to time) differ among type 1 and 2 and TD antigens. The response to type 1 and 2 antigens consists mainly of IgM. The response may persist for several months with the concentration of antibody in the serum fluctuating up and down with a period of 1–2 weeks. This cyclical phenomenon probably results from initial clearance of antigen by macrophages (which may ingest it, but may not have the enzymes necessary to degrade it), leading to an initial decrease in antibody production, followed by a re-release of antigen and renewed antibody synthesis. Macrophages lack enzymes capable of degrading the majority of T-independent antigens of bacterial or plant origin.

In the primary response to a TD antigen, there is a logarithmic increase in serum IgM antibodies from 4–10 days. This increase reflects clonal expansion and antibody production. Clonal expansion is associated with class-switching so that the proportion of IgG antibody increases as time goes on. Recall that B cells may develop into either plasma cells or memory cells. The latter are morphologically similar to small resting lymphocytes, but differ in important ways. Memory cells circulate more actively from blood to lymph, and they survive for much longer periods of time between mitoses.

The secondary humoral response is characterized by a much shorter lag period, and predominance of IgG. This rapidly-appearing IgG component is derived from memory cells developed in the primary response. Memory cell Ig receptors have higher affinity for the antigen, as do the antibodies secreted by their progeny in the secondary response. For this reason, they are stimulated by much lower antigen concentrations than are virgin B cells. In addition, memory cells are more receptive to T cell help. The signals leading to development of memory cells versus plasma cells are unknown, as are details of the memory cell's activation (apart from having Ig receptors of higher affinity) which give it such an advantage over virgin B cells.

The rapidity of the secondary response is the basis of immunity acquired by either natural infection or vaccination. While the primary response may be too slow to protect against certain pathogens, the secondary response often is fast enough to halt progress of infections. Vaccination is one of the most important contributions of immunology to disease prevention.

Regulation of the antibody response

We have examined the mechanisms responsible for the positive control of the humoral response. That is, those mechanisms which act to activate B cells and induce antibody secretion. Most prominent among these is the action of helper T cells which are critical for production of antibodies to the majority of protein and glycoprotein antigens. What mechanisms act to suppress the humoral response and bring it to an end when it is no longer required?

Antibodies themselves act to reduce stimulation of the humoral response by several mechanisms. The first is by *clearance* of antigen. As the concentration of serum antibody increases, immune complexes begin to form. Cells of the mononuclear phagocyte system (see Chapter 2), predominantly macrophages in the spleen and other tissues, possess receptors for Ig Fc and bind immune complexes, ingest and degrade them. As the amount of circulating antigen falls, there is less available for stimulating antibody secretion.

Antigen circulating in a complex with antibody is itself much less effective in initiating a humoral response. Hence, even small immune complexes which are not as rapidly cleared as larger ones do not induce B cell activation. Furthermore, immune complexes themselves may interact with B cell Ig and Fc receptors and render B cells inactive.

Another mechanism, whose importance is debated, involves the activity of a specialized class of T cells, *suppressor cells*. These will be discussed in Chapter 7.

Lastly, we present another mechanism which has been proposed as a component in the regulation of antibody responses: the idiotype network.

Idiotype network theory

In 1974 Niels Jerne presented the idiotype network theory. This hypothesis was motivated by the discovery that antibodies reacting with other antibodies occurred spontaneously during immune responses and could be induced artificially by immunization with antigens or antibodies. These antibody antigenic determinants recognized by other self antibodies were found to be associated with the variable region and were called idiotypes (see Chapter 5). An antibody whose combining site (paratope) interacts with an antigenic determinant on another antibody variable region (idiotope) is called an *anti-idiotypic antibody.* (For convenience, the idiotype-bearing antibody is often abbreviated Ab_1, while the anti-idiotype antibody is Ab_2.) Most simply stated, the Ab_1 produced in response to an antigen themselves elicit a humoral response (Ab_2) directed against Ab_1 idiotopes. Ab_2 then acts to suppress the synthesis of Ab_1. T cells also recognize idiotopes and may play a role in idiotype-mediated regulation of the immune response (more of this in Chapter 7).

It has been repeatedly demonstrated that idiotypes (Ab_1) and anti-idiotypes (Ab_2) may suppress or stimulate production of particular idiotypes or antigen specificities associated with them. Many studies have shown that T cell antigen receptors both express and recognize idiotopes. Furthermore, it has been observed that injection of Ab_2 into pregnant mice may enhance or suppress expression of Ab_1 in the offspring. Thus, maternal antibodies transmitted to the fetus across the placenta, or to the neonate in milk, influence the early repertoire and may establish a lifelong pattern via idiotypic interactions. Although it is well-established that idiotype-anti-idiotype interactions occur in immune responses, their role as regulatory forces remains a subject of much controversy.

Several important phenomena with practical applications have been described in the course of investigations of idiotypes. One has been the exploitation of idiotypes as markers of B cell neoplasias. A B cell tumor which has surface antibody is subject, in some instances, to regulatory signals transmitted by them. Anti-idiotypic antibodies may slow the growth of neoplastic B cells, and some remissions of B cell lymphoma following anti-idiotype treatment have been reported.

ABNORMALITIES OF B CELL DEVELOPMENT

Derangements of B cell differentiation may occur as a result of genetic defects, or of neoplastic transformation of a B cell resulting in a malignant B cell clone. Genetic defects are characterized by variable immunodeficiency due to decreased or absent levels of antibodies of some or all classes. Malignant transformation of a B cell results in the proliferation of a single clone which may produce large quantities of a single antibody. The latter syndromes are referred to as *monoclonal gammopathies*, and are also characterized by variable immunodeficiency.

Genetic defects

X-linked agammaglobulinemia (XLA)

Limited to males, this disease manifests itself 6 to 10 months after birth, and is characterized by recurrent pyogenic infections (*S. pyogenes*, *S. aureus*, *H. influenzae*). Serum immunoglobulins are low or absent; B lymphocytes are rare or absent, and plasma cells are not found. The XLA gene has been mapped to the proximal portion of the long arm of the X chromosome. Patients have pre-B cells or IgM^+-IgD^+ B cells in the bone marrow, but the immunoglobulin is aberrant, consisting only of a leader sequence joined to DJ_HC. Apparently, the gene involved in this disease controls some aspect of immunoglobulin gene rearrangement, transcription, or RNA splicing. This supposition is strengthened by recent experiments showing that murine myeloma cells are able to complement the X-linked defect in the B cells from these patients. Hybridomas made by fusing B cells from an XLA patient with only D-J rearranged, with a murine myeloma, went on to rearrange V-DJ. The expressed V gene is of human origin. The rearrangement mechanism supplied by the murine myeloma complemented the defect in the XLA cell.

Common variable hypogammaglobulinemia

This syndrome (also called *common variable immunodeficiency*) occurs later in life (the second and third decades) and may have varying manifestations. Patients usually have peripheral blood B cells, but few plasma cells, and low levels of antibodies. At least three different etiologies are grouped under the name common variable hypogammaglobulinemia: a predominant B cell defect; a predominant T cell defect, either an increase in T suppressors or a decrease in T helpers; or the production of autoantibodies reacting with B and/or T cells.

Monoclonal gammopathies

There are three major monoclonal gammopathies: *multiple myeloma*, *Waldenström's macroglobulinemia*, and the *heavy chain diseases*. These conditions are also frequently referred to as *plasma cell dyscrasias*.

Multiple myeloma

Characteristic of this disease, as in most gammopathies, is appearance in the serum of a large amount of homogeneous protein. When observed in immunoelectrophoresis, this protein peak is called the *M-component* or *M-spike* (M = myeloma). The M component is immunoglobulin secreted by a malignant B cell clone. The urine of the majority of patients contains *Bence-Jones proteins*, excess Ig light chains produced by the myeloma and filtered by the kidney.

Multiple myeloma patients may exhibit multifocal bone pain and pathological fractures, renal failure, normochromic normocytic anemia, and hypercalcemia. The skeletal symptomatology, anemia, and hypercalcemia result from proliferation of malignant cells in the bone marrow and bone resorption. Diagnosis is usually made by detecting the M component by serum protein electrophoresis in association with osteolytic lesions seen in radiographs. Afflicted individuals have impaired ability to generate humoral immune responses.

Antibodies produced by myeloma cells are called *myeloma proteins*. Roughly 54% are IgG, 23% IgA, 1% IgD, and about 20% consist of light chains only. Other isotypes are rare. About 1% of myelomas are non-secretory. Although the clinical manifestations of multiple myeloma are due to accumulation of mature plasma cells, the malignant transformation appears to occur at the pre-B cell stage, and allows the transformed progeny to differentiate into plasma cells independently of external regulatory signals.

Myeloma protein idiotypes function as *tumor-associated antigens* since they are usually not found in high amounts on normal lymphocytes. Myeloma cell growth may be affected by passively administered anti-idiotype antibodies, and anti-idiotypic immunity may protect animals from lethal doses of myeloma cells. Myeloma cells have also been shown to respond to signals delivered by helper and suppressor T cells.

Waldenström's macroglobulinemia

This gammopathy results from proliferation in the bone marrow of IgM-secreting plasmacytoid cells. The clinical presentation is non-specific, most often generalized weakness, fatigue, and weight loss. Diagnosis is made by detecting the M component by serum protein electrophoresis. Common associated symptoms are anemia and hepato-splenomegaly. Bence-Jones proteinuria occurs in about 10% of patients. Osteolytic bone lesions are very rare. The disease is relatively indolent, and even though incidence is rarely before age 60, the mean survival after diagnosis is three to five years, and occasionally ten years or more. Death attributable to this disease is usually due to increased blood viscosity. This can result in renal failure, hemorrhagic purpura of mucous membranes (paraproteinemia may interfere with coagulation), and occasionally congestive heart failure.

Heavy chain diseases

The rare *heavy chain diseases* appear secondarily to a primary malignancy of lymphoid tissue. The malignant cells have lost the ability to synthesize a complete antibody molecule, and produce only the heavy chain or a part of it.

Gamma-chain disease, also known as *Franklin disease*, manifests clinically as malignant lymphoma with lymphadenopathy and hepato-splenomegaly. Proteinuria is common. While the serum paraprotein is a γ-chain fragment, the urinary protein is most frequently a κ light chain. There is increased susceptibility to bacterial infection which is usually the proximate cause of death. The paraproteins are usually not complete antibodies. For example, the expressed heavy chain genes cloned from a patient's cells showed a normal leader sequence with four V_H amino acids joined to the last amino acid encoded by a J gene. This indicates a deletion of the largest V_H exon, the D gene, and most of J.

Alpha-chain disease, also called *Seligmann disease*, is associated with lymphomas involving the gastrointestinal system, the major area of IgA synthesis. Symptoms include malabsorption, diarrhea, and extreme weight loss.

The clinical presentation of *μ-chain disease* is that of chronic lymphocytic leukemia (lymphocytosis, hepatosplenomegaly, and amyloidosis). In the few reported cases of this disease, the bone marrow contains abnormal vacuolated plasma cells. The major serum paraprotein is a pentameric μ-chain fragment.

Selective immunoglobulin class deficiencies

In these diseases, called *dysgammaglobulinemias*, absence or low levels of one or more Ig classes results in variable immunodeficiency. The most common dysgammaglobulinemia is *IgA deficiency*, observed at a frequency of 1:3–400 among blood donors, and is most often asymptomatic. Usually both IgA isotypes (IgA1 and IgA2) are absent, but selective deficiency of one IgA isotype has been observed. The serum level of other isotypes is most often normal. Symptomatic individuals often have recurrent respiratory, gastrointestinal, and urinary tract infections. Some may develop symptoms similar to sprue. IgA deficiency may be associated with autoimmune diseases such as systemic lupus erythematosus. The basis of this association is unclear.

Selective deficiencies of other isotypes are extremely rare. *IgG2 deficiency* has been observed most often in patients with IgA deficiency. *IgM deficiency* has been associated with recurrent pyogenic infections and gram-negative septicemia. Deficiency of λ light chains has also been reported.

Dysgammaglobulinemias may have more than one etiology. In most cases examined, the individual still possesses the gene encoding the isotype in which they are deficient. In these cases, the defect is probably in the T cell control of isotype expression (see Chapter 7). However, in a few patients with selective IgA1 deficiency, absence of the $\alpha 1$ gene was observed. One case has been reported in which the patient was unable to synthesize secretory component and could not produce secretory IgA. This individual presented with chronic intestinal candidiasis.

As mentioned several times in this and preceding chapters, B cells in isolation are not able to generate antibodies in response to many antigens. They require help from another group of lymphocytes, the T cells. Not only do T cells regulate the activity of immunocompetent cells (including themselves), but they are also effectors of another form of specific immunity: the cell-mediated response. These cells are the subject of the next chapter.

SOURCES AND SUGGESTED ADDITIONAL READING

Bona, C. A. (1987) *Regulatory Idiotopes*, John Wiley & Sons, New York.

Buckley, R. H. (1986) Humoral immunodeficiency. *Clin. Immunol. Immunopathol.*, **40**:13–24.

Cambier, J. C., Justement, L. B., Newell, M. K., Chen, Z. Z., Harris, L. K., Sandoval, V. M., Klemsz, M. J. & Ransom, J. T. (1987) Transmembrane signals and intracellular "second messengers" in the regulation of quiescent B-lymphocyte activation. *Immunol. Rev.*, **95**:37–57.

Casali, P. & Notkins, A. L. (1989) Probing the human B-cell repertoire with EBV: polyreactive antibodies and $CD5^+$ B lymphocytes. *Annu. Rev. Immunol.*, **7**:513–535.

Coffman, R. L., Seymour, B. W. P., Lebman, D. A., Hiraki, D. D., Christiansen, J. A., Shrader, B., Cherwinski, H. M., Savelkoul, H. F. J., Finkelman, F. D., Bond, M. W. & Mosmann, T. R. (1988) The role of helper T cell products in mouse B cell differentiation and isotype regulation. *Immunol. Rev.*, **102**:5–28.

Houghton, A. & Scheinberg, D. (1988) Monoclonal antibodies in the treatment of hematopoietic malignancies. *Semin. Hematol.*, **25** (Suppl. 3):23–29.

Jerne, N. K. (1974) Towards a network theory of the immune system. *Ann. Immunol. (Inst. Pasteur)*, **125C**:373–389.

Langman, R. E. & Cohn, M. (1986) The 'complete' idiotype network is an absurd immune system. *Immunol. Today*, **7**:100–101.

MacLenna, I. C. & Gray, D. (1986) Antigen-driven selection of virgin and memory B cells. *Immunol. Rev.*, **91**:61–85.

Morgan, E. L., Hobbs, M. V., Thoman, M. T. & Weigle, W. O. (1986) Lymphocyte activation by the Fc region of immunoglobulins. *Immunol. Invest.*, **15**:625–687.

Osserman, E. F., Merlini, G. & Butler, V. P., Jr. (1987) Multiple myeloma and related plasma cell dyscrasias. *JAMA*, **258**:2930–2937.

Pike, B. L., Alderson, M. R. & Nossal, G. J. V. (1987) T independent activation of single B cells: analysis of overlapping stages in the activation pathway. *Immunol. Rev.*, **99**:119–152.

Schwaber, J., Koenig, N. & Girard, J. (1988) Correction of the molecular defect in B lymphocytes from X-linked agammaglobulinemia by cell fusion. *J. Clin. Invest.*, **82**:1471–1476.

Weigle, W. O. (1987) Factors and events in the activation, proliferation, and differentiation of B cells. *CRC Crit. Rev. Immunol.*, **7**:285–324.

Chapter 7

T Cells and Cellular Immunity

T lymphocytes are so-named because they mature in the thymus. T cells mediate cellular immunity, and regulate the proliferation and activity of B cells, T cells, and other cells participating in immune responses (accessory cells). Cellular immunity is important in the elimination of cells infected with intracellular pathogens (e.g., viruses) and cells exhibiting aberrant cytodifferentiation (neoplastic cells). Cellular immunity also operates in the destruction of allogeneic cells (graft rejection).

T CELL DEVELOPMENT

T cell development is divided into three stages, the *prethymic phase*, the *thymic phase*, and the *mature phase* (Figure 7.1).

The prethymic phase

The prethymic phase, during which pluripotent stem cells develop into *prethymic cells* or *pre-T cells*, begins with hemopoiesis in the fetal liver and spleen, then in the bone marrow later in gestation and thereafter. Prethymic cells reach the thymus via blood, enter by passing between endothelial cells in venules, and find their way to the outer cortex.

Several cell-surface molecules serve as markers of T cell differentiation. The earliest marker identified in T cell development is CD7. This early stage is not yet committed to the T cell lineage and may enter erythroid, myeloid, or megakaryocytoid pathways. Nevertheless, CD7 expression appears to be important for T cell development since its

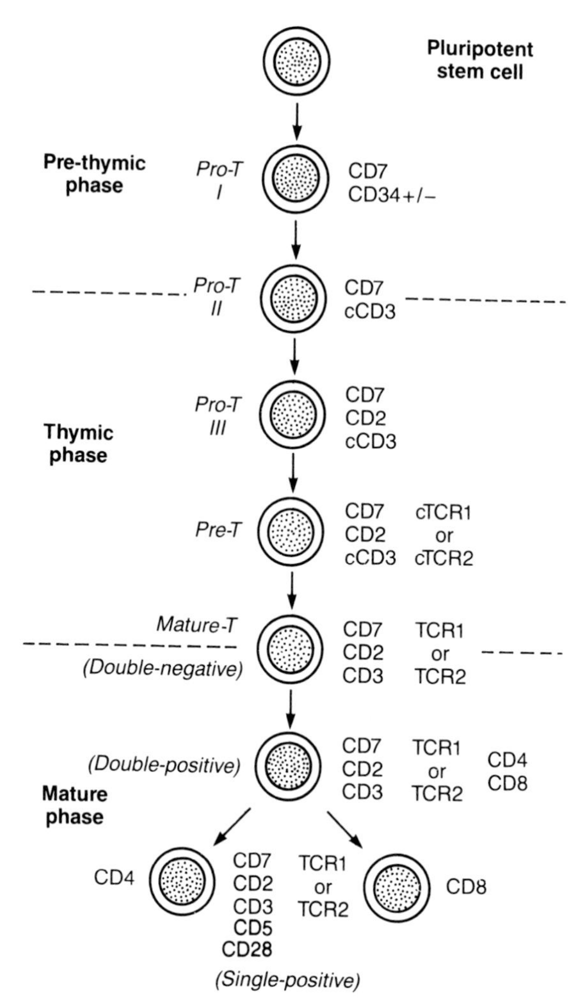

Figure 7.1. *T cell development.* The earliest defined T cell marker is CD7 (the pro-T I stage). Not all $CD7^+$ cells become T cells. Some early T cells also transiently express CD34. The appearance of cytoplasmic CD3 (cCD3) marks transition to the pro-T II stage. This is the cell which then enters the thymus to continue development. Appearance of the adhesion molecule CD2 marks the pro-T III stage. The pre-T cell is distinguished by cytoplasmic TCR proteins either TCR1 (γ/δ) or TCR2 (α/β). When these receptors appear on the surface, the cells are designated mature T cells. These are called *double-negative cells* since they express neither CD4 nor CD8. T cells then acquire both CD4 and CD8, becoming *double-positive cells.* The T cell then goes on to express only CD4 ($T_{h/i}$ phenotype) or CD8 (T_c or T_s phenotype). These are also called *single-positive cells.* At some time within the thymus, T cells also acquire the CD5 and CD28 surface molecules. (Adapted in part from Haynes et al., 1989.)

absence from pre-T cells has been associated with severe immunodeficiency. The role of CD7 in T cell differentiation is unknown. Some $CD7^+$ cells also express CD34. Both $CD7^+/CD34^+$ and $CD7^+/CD34^-$ cells may become either T cells or other cell types. The function of CD34 is unknown. The next CD antigen to be expressed is CD3. Although CD3 is destined to appear on the cell surface where it has an important association with the T cell antigen receptor, its expression in early prethymocytes is only in the cytoplasm.

The thymic phase

At the beginning of the thymic phase, cells are dividing rapidly and have the appearance of lymphoblasts. After this proliferative phase, cells take on the appearance of small lymphocytes and migrate through the cortex toward the medulla. Developing T cells within the thymus are called *thymocytes*. (The histological organization of the thymus is described in Chapter 2.)

Several types of specialized thymic epithelial cells play critical roles in T cell development. These cells produce several *thymic hormones* required for thymocyte differentiation. Three that have been partially characterized are *thymopoietin*, a 7 kd protein; *thymosin*, a 12 kd protein; and *thymosterin*, a steroid. Another factor, interleukin 7 (see below) is also produced by thymic epithelial cells and promotes cell division in immature thymocytes.

Thymic epithelial cells are also important in establishing the *repertoire* of antigen specificities of mature T cells. The concept of the T cell repertoire is identical to that of the B cell repertoire. However, due to the very different ways in which B cells and T cells interact with antigen, the set of epitopes recognized by each cell type is different (although not necessarily mutually exclusive, some epitopes may be recognized by both). In order for the immune system to be effective, the T cell repertoire must be large enough to recognize the multitude of foreign molecules the body will encounter. At the same time, these cells must not direct their activity toward our own epitopes. How does this come about?

Thymic epithelial cells express on their surfaces glycoproteins encoded by genes of the *major histocompatibility complex* (MHC, the subject of Chapter 8). Unlike immunoglobulins, which bind to immunogens in their native state, T cell antigen receptors bind only antigen fragments that are complexed with MHC molecules.

When lethally irradiated mice of strain "a" are reconstituted with bone marrow cells from mice of strain "b", the T cells of the resulting chimeras recognize antigens in association with strain "a" histocompatibility antigens. That is, T cells function in the context of the MHC proteins encountered in the thymus in which they matured, not those encoded within their own genomes, or those present in their environment during the prethymic phase of development.

This "education" of thymocytes (formation of the T cell repertoire) is not the modification of the activity or surface characteristics of individual cells. Rather, it is the *selection* of a group of cells which recognize non-self antigens, and do not respond to self epitopes. The outcome of the selection process is severe. Only 1–5% of prethymocytes exit the thymus as mature T cells, the remainder are destroyed.

T cell selection in the thymus is hypothesized to occur by both positive and negative mechanisms. As mentioned above, T cells recognize peptides associated with self MHC molecules. As a first requirement, then, mature T cells must be able to interact to some extent with self MHC proteins. Positive selection is the preservation of these cells, and the death of cells not interacting with self MHC molecules. The second requirement is that mature T cells not react with self peptides associated with self MHC proteins. Negative selection is the destruction of those cells that do.

Foreign antigens are not found in normal thymuses. How, then, is specificity for foreign epitopes selected for? Several hypotheses have been proposed. Recent studies have suggested that in addition to the MHC molecules found throughout the body, thymic epithelia may express additional MHC proteins restricted to the thymus. These specialized MHC molecules may permit selection of T cells which will recognize foreign peptides associated with the MHC molecules in the periphery. In another model, thymic epithelia actually generate "foreign" peptides through errors in transcription and translation. Aberrant peptides resulting from the misincorporation of nucleotides in mRNA, and of amino acids in proteins, could be sufficiently diverse to establish an adequate repertoire. After leaving the thymus, the likelihood of a T cell encountering the same "accidental" peptide, and initiating a response in the absence of a foreign source of antigen, is infinitesimal.

The most immature thymocytes are $CD7^+$ and have cytoplasmic CD3. The next marker to be expressed is CD2. CD2 is an *adhesion molecule*, it is important in direct cellular contacts between T cells and other cells participating in immune responses (see below). Subsequently, thymocytes express CD1 (unknown function), CD3 (on the surface now), CD4, CD5, CD8, CD28, CD45 and one of two types of antigen receptor (the T cell receptor or TCR). In an intermediate stage of development, thymocytes are $CD3^+/CD4^-/CD8^-$, these are called *double-negative* cells. Subsequently they become $CD3^+/CD4^+/CD8^+$, or *double-positive* cells. At this point, the cell then loses either CD4 or CD8 becoming $CD3^+/CD8^+$ or $CD3^+/CD4^+$ (*single-positive*). CD4 and CD8 distinguish different subtypes of T cells (see below).

The mature phase

Mature thymocytes exit the thymus through venule walls in the medulla and the corticomedullary junction and populate peripheral lymphoid organs. These cells may have a long life span, perhaps the entire life of the individual. Upon antigen stimulation, T cells undergo rapid proliferation and take on a blastoid appearance, as do B cells. After stimulation, T cells become *effector cells* and some types of T cell may also become memory cells analogous to memory B cells.

Characteristics of mature T cells

Mature T cells are small lymphocytes morphologically indistinguishable from other lymphocyte subpopulations. However, as for B cells, there are several cell-surface and metabolic markers specific for T cells and T cell subclasses.

The enzyme *terminal deoxyribonucleotidyl transferase* (*TdT*) is a template-independent DNA polymerase that has been found in relatively high amounts in thymocytes of all species. This enzyme is also found in B cells.

Prethymic, and 95% of thymic T cells are highly sensitive to *cortisone* and are destroyed when it is administered in large doses. Mature peripheral T cells and 5% of thymic T cells (the most mature cells in the cortex) are resistant to cortisone.

As do B cells, T cells respond with varying degrees of proliferation and activation to several lectins of plant origin (*T cell mitogens*). Two of the most widely used in clinical studies of T cell function are concanavalin A (Con A), and phytohemagglutinin (PHA) derived from different types of beans.

T cell subsets

All mature thymocytes and peripheral blood T cells bear the CD2, CD3, CD5, and CD28 antigens. Other surface molecules serve to differentiate several functionally distinct subpopulations of mature T cells. These subsets differ in their interaction with various classes of self histocompatibility proteins, and their secretion of various factors (cytokines) influencing proliferation and differentiation of other lymphocytes. Two principal categories of T cells have been defined, effector cells mediating cellular immune responses, and regulatory cells modulating the activity of B cells and other T cells.

Effector T cells are further subdivided. *Cytotoxic T cells* (T_C) express CD8, and lyse autologous cells bearing foreign antigen molecules associated with class I histocompatibility proteins, or cells allogeneic with respect to MHC class I. Cells mediating the *delayed type hypersensitivity* (*DTH*) reaction (T_{DTH}) express CD4 and recognize foreign antigens in association with class II histocompatibility molecules. *Contact sensitivity* (*CS*) is a special case of DTH in which the antigens are skin proteins having undergone adventitious chemical modification. The CS reaction may be mediated by a population of T cells (T_{CS}) distinct from T_{DTH}, but this has not been firmly established.

Regulatory T cells are divided into two major subgroups. One is the *helper* or *inducer T cell* (T_h or T_i) which expresses CD4 and secretes factors inducing proliferation and differentiation of B cells and other T cell classes. Some authors refer to all T cells which enhance proliferation and activity of other immune cells as T_h, regardless of whether their targets are T cells or B cells. Others may define T_i cells as activators of other T cells (cytotoxic, suppressor, or helper), and T_h cells as those cooperating with B cells in humoral immune responses. Both types of cells are often referred to as helper/inducers or $T_{h/i}$. $CD4^+$ T cells are functionally heterogeneous, but we do not yet have sufficient information to categorize them all systematically. We will henceforth refer to the cells providing help to B cells as T_h, and those activating other T cells as T_i. $T_{h/i}$ will designate both groups of $CD4^+$ cells.

In mice, the helper T cells which activate B cells in humoral responses have been divided into two subtypes designated T_h1 and T_h2. These cells differ in the types of cytokines they secrete, the antibody isotypes they induce, and the types of immune responses in which they participate. While human T_h cells also appear to be heterogeneous, it is not yet clear that they can be classified in a manner analogous to those of mice.

The other principal type of regulatory T cell is the *suppressor T cell* (T_s). This class of cells expresses CD8 and inhibits the activation of B or T cells. The suppressor T cell has been the subject of much controversy in immunology over the past decade. Some aspects of suppressor cell function will be discussed below.

THE T CELL ANTIGEN RECEPTOR (TCR)

T cells are clonally distributed analogously to B cells; their clonality manifested most clearly in the antigen receptor. As are immunoglobulins, T cell receptors are composed of two polypeptide chains; each chain having constant and variable regions; expressing idiotypic determinants associated with variable regions; and encoded by multiple gene segments, diversity arising from recombination of these segments. As mentioned above, however, a most important distinction between immunoglobulin receptors and TCRs is that the former have evolved to bind to antigen alone, while the latter bind to antigen fragments associated with cell-surface histocompatibility molecules.

Two distinct types of TCR have been characterized. Mature cytotoxic, suppressor, and helper/inducer T cells bear a receptor made of two chains called α and β. This heterodimer is intimately associated with the CD3 molecule in the cell membrane (Figure 7.2). The α chain is a glycoprotein of $M_r = 40{,}000$ while the β chain glycoprotein has $M_r = 50{,}000$. Figures 7.3 and 7.4 show the organization of TCR α and β genes, respectively. The CD3 antigen is a complex of several different

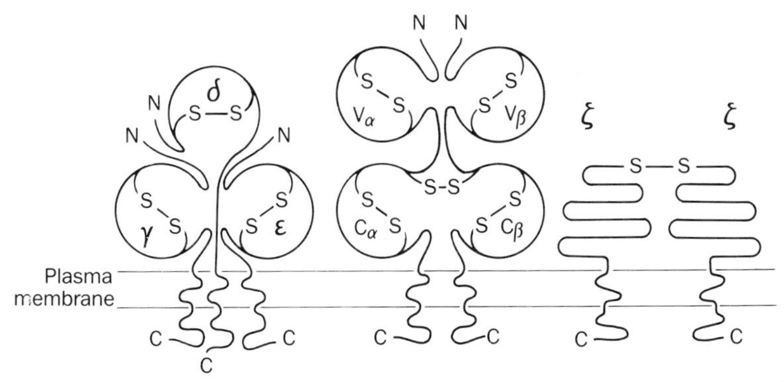

Figure 7.2. *Structure of the T cell receptor-CD3 complex.* The T cell antigen receptor itself is comprised of two chains, α and β, (or γ and δ) while CD3 consists of five polypeptides, one each of γ, δ (different from TCR γ and δ), and ε, and two of ζ (zeta). The carboxyl termini of these proteins are intracytoplasmic. These polypeptides are intimately associated in the membrane, their precise physical relationship one to another is not known.

polypeptides. This complex appears to function in transducing a signal of TCR occupancy across the membrane and initiating or contributing to T cell activation. The α/β TCR is also called *TCR2*.

Another distinct subset of T cells bear a different type of TCR made of γ and δ glycoproteins. The γ chain has $M_r = 40{,}000$. Three different types of δ chains have been described. One has $M_r = 40{,}000$ and is disulfide-linked to the γ chain in the receptor. Two others have $M_r = 40{,}000$ and 55,000 and are not disulfide linked with γ. One 40 kd form is encoded by the $C_\delta 1$ gene, the 50 kd form by the $C_\delta 2$ gene (Figure 7.5). The gene encoding the other 40 kd form has not yet been located. The significance of these different types of TCR with respect to functional characteristics of the cells expressing them is unknown. As is the α/β TCR, the γ/δ TCR is intimately associated with the CD3 molecule in the plasma membrane. The γ/δ TCR is also known as *TCR1*.

Recently, a leukemic T cell line expressing receptors consisting of β/δ heterodimers has been identified. Whether this represents another functional form of TCR or is simply an aberration remains to be seen.

Studies in mice show that $TCR1^+$ cells predominate early in ontogeny, and subsequently decrease markedly in number, being replaced by $TCR2^+$ cells. T cells bearing TCR1 constitute only 1–2% of the total population in adults. Furthermore, most are double-negative cells ($CD3^+/CD4^-/CD8^-$). Recently, it has been shown that a $TCR1^+/CD8^+$ T cell line from an individual immunized with tetanus toxoid recognized antigen *in vitro* in association with MHC in a manner entirely analogous to $TCR2^+$ T cells. The functional consequences of expressing TCR1 versus TCR2 are not known. The physiological role of $TCR1^+$ double-negative cells is also unclear.

The variable regions of TCR proteins are divided into framework and hypervariable regions, as are antibody V regions. TCRs also exhibit allotypic and idiotypic determinants. T cell antigenic markers are usually identified with antibodies from animals immunized with T cell clones. However, TCR (and antibody) epitopes may also be recognized by other TCRs. Idiotype or allotype specific antibodies and T cells may activate or suppress the expansion of T or B cells exhibiting the determinant.

Genes encoding the TCR

The chromosomal location of the various TCR gene loci is summarized in Table 7.I.

Table 7.I. CHROMOSOMAL LOCATIONS OF T CELL RECEPTOR GENES

Locus	Chromosome
α,δ	14
β,γ	7

The α locus, analogous to immunoglobulin light chains, contains three gene clusters, V_α, J_α, and C_α (Figure 7.3). As is the immunoglobulin heavy chain, the TCR β chain is encoded by four gene segments. Figure 7.4 shows the organization of the β chain gene clusters, V_β, and a tandem of $D_\beta 1$, $J_\beta 1$, $C_\beta 1$, and $D_\beta 2$, $J_\beta 2$, $C_\beta 2$. The 3′ flanking region of V_β genes contain heptamer and nonamer sequences separated by a 23 base pair spacer. D_β genes are also flanked by nonamer and heptamer sequences with 12 bp spacers on either side, while J_β genes have flanking sequences with 23 bp spacers on the 5′ side. Gene recombination follows the 12–23 base pair rule (see Chapter 6), suggesting that TCR genes rearrange via a mechanism very similar, if not identical, to that for immunoglobulin genes.

The overall structure of the γ/δ TCR is identical to that of the α/β TCR (Figure 7.2). The organization of the γ and δ loci is shown in Figures 7.5 and 7.6, respectively. The δ locus is positioned between V_α and J_α (Figure 7.6). Thus, a cell which has rearranged the α locus no longer contains DNA encoding the δ chain.

Just as with immunoglobulins, TCR diversity arises from gene recombination, imprecise joining of gene segments, and the insertion of nucleotides at the gene joints (N regions). However, somatic mutation does not occur in TCR genes.

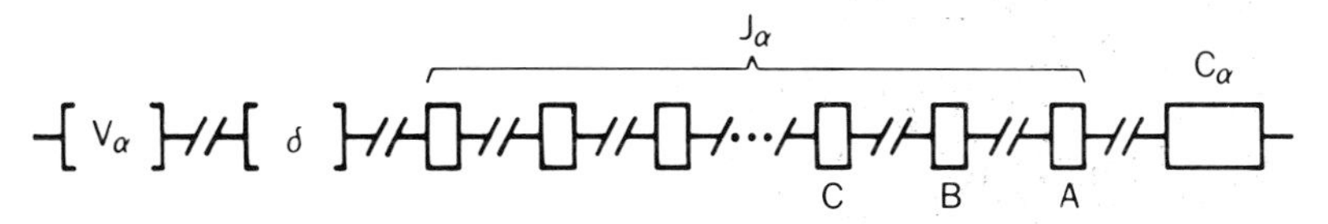

Figure 7.3. *Organization of the human T cell receptor α chain locus.* Estimates of the number of V_α genes range from 20–100. The number of J_α genes is thought to be on the order of 50. J_α genes are designed A, B, C, etc. beginning with the most 3′ (nearest to C_α. There is only one C_α gene. Rearrangement of α genes is analogous to that of immunoglobulin light chains (Figure 5.2). Note the position of the δ locus between V_α and J_α. (Adapted from Toyonaga and Mak, 1987.)

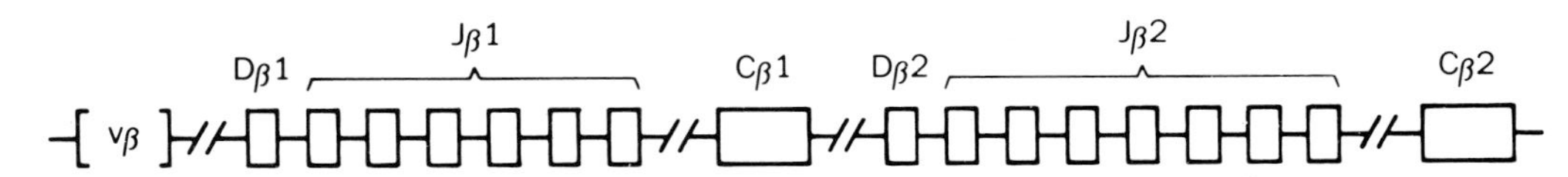

Figure 7.4. *Organization of the human T cell receptor β chain locus.* The number of V_β genes is unknown, estimates range from 70 to several hundred. D_β, J_β, and C_β genes are arranged in two clusters, 1 and 2. Rearrangement of these genes is analogous to that of immunoglobulin heavy chains (Figure 5.3). (Adapted from Toyonaga and Mak, 1987.)

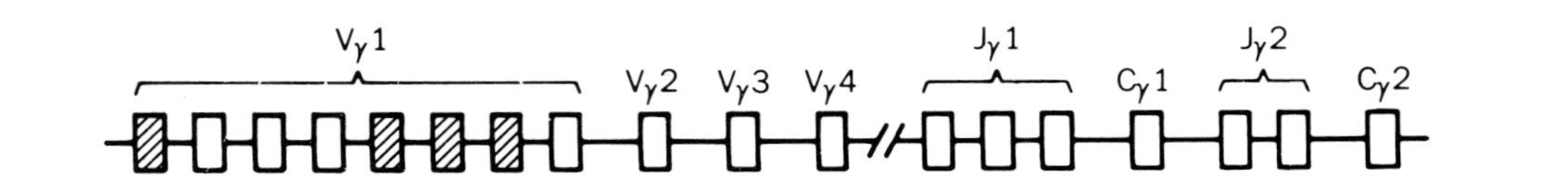

Figure 7.5. *Organization of the human T cell receptor γ chain locus.* The γ chain variable region genes are divided into four families $V_\gamma1$–4, $V_\gamma1$ contains 8 genes, four are pseudogenes (shaded). $V_\gamma2$–4 each have only one member. There are two J_γ-C_γ clusters. Rearrangement is as with immunoglobulin light chains (Figure 5.2). (Adapted from Strauss et al., 1987.)

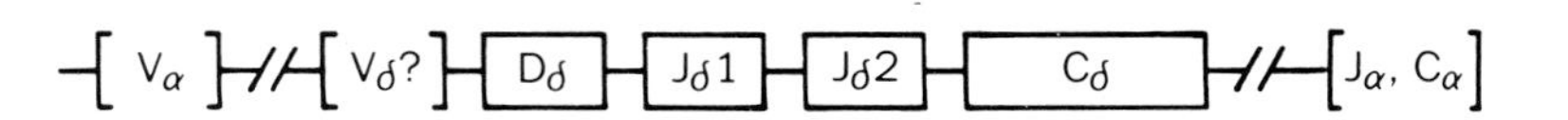

Figure 7.6. *Organization of the human T cell receptor δ chain locus.* The number of V_δ genes is not known, but is thought to be quite small. There is only one D_δ gene, two J_δ genes, and one C_δ gene. Note the position of the δ locus between V_α and J_α. Rearrangement is as with immunoglobulin heavy chain genes (Figure 5.3). (Adapted from Griesser et al., 1988.)

ANTIGEN PRESENTATION

The majority of antigens encountered in nature are not, by themselves, capable of activating T or B cells. Activation requires action of *accessory cells* or *antigen-presenting cells* (*APCs*). Classically, this function has been ascribed to cells derived from monocytes, including macrophages and dendritic cells present in various tissues. Antigen presentation is the association of antigen, or a part of it, with APC MHC molecules. Two pathways of antigen presentation may be distinguished depending on the type of T cell which is being activated. Recall that helper/inducer T cells recognize antigen in association with self MHC class II molecules, while cytotoxic T cells require class I MHC molecules for antigen recognition.

Antigen presentation to class II-restricted cells involves antigen *processing* by the APC. Many studies indicate strongly that internalization and partial degradation of antigen by APCs are important for presentation. Antigen fragments become associated with MHC molecules within phagolysosomes and are translocated to the cell surface (Figure 7.7). Macrophage presentation of intact protein antigens is inhibited by drugs such as chloroquine, which perturbs lysosome function, or protease inhibitors. Glutaraldehyde-fixed macrophages, and liposomes (artificially-produced membranous vesicles) containing class II antigens may effectively present small peptides, but not intact proteins. It is not clear how dendritic cells, which have low or absent phagocytic capacity, and a low content of degradative enzymes such as proteases and esterases, in many instances function very well as antigen presenting cells. Recent research indicates that B cells also function as APCs. Apparently, B cells bind antigen via immunoglobulin receptors, internalize and process it.

Perhaps these different processing mechanisms correspond to characteristics of the antigens themselves, such that some require internalization and digestion while others do not. It is generally accepted that most native proteins require some intracellular processing for presentation, while smaller peptides or denatured proteins may not.

The final step in presentation to $T_{h/i}$ cells is association of antigen with MHC class II molecules. This complex is the unit recognized by the antigen receptor of the T cell being activated. Different antigens or even different fragments of the same antigen may complex to varying degrees with the class II molecules encoded by different MHC alleles. This is the basis of *immune response gene* phenomena to be described in Chapter 8. Remember also that in order for antigen presentation to be

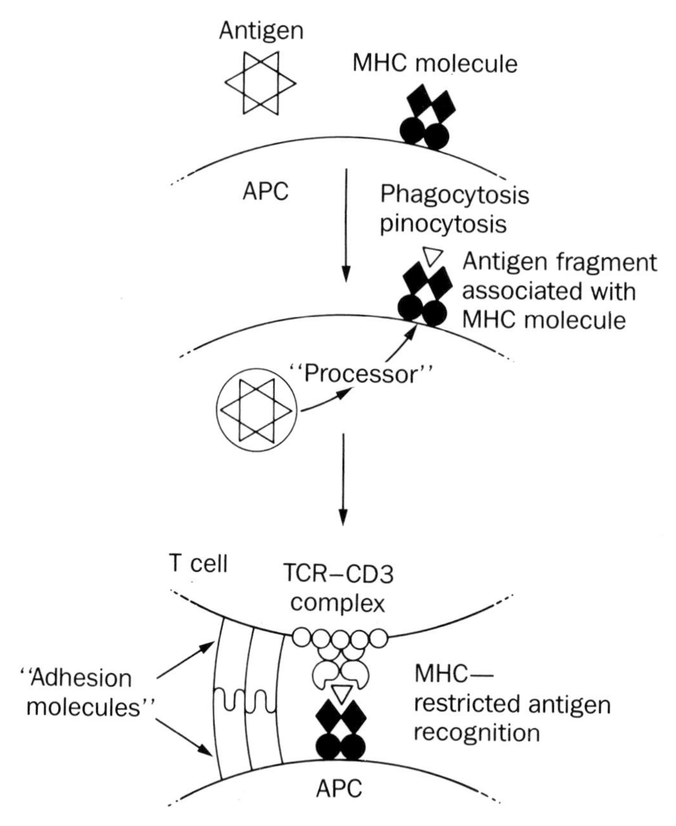

Figure 7.7. *Antigen presentation.* Two pathways of antigen presentation may be distinguished depending on the type of T cell that is being activated. Helper/inducer T cells recognize antigen in association with MHC class II molecules. Antigen presented to these cells is internalized, partially degraded (processed) and returned to the cell surface complexed with class II molecules (as shown in the first two diagrams here). Cytotoxic T cells recognize antigen in association with MHC class I molecules. Antigen presented to these cells is synthesized *de novo* within the presenting cell (e.g., during a viral infection) and subsequently expressed on the cell surface complexed with class I molecules. The actual recognition step involves binding of the T cell TCR-CD3 complex to the antigen-MHC complex (last diagram). This interaction is stabilized by adhesion molecules on the T cell and antigen-presenting cell (APC) surfaces. $T_{h/i}$ CD4 and T_c CD8 bind MHC class II and class I determinants, respectively. T cell CD2 also binds APC CD58. Monocyte-derived cells and dendritic cells in skin and lymphoid tissues are the predominant APCs for T_h cells. B cells also function as APCs, primarily of the antigens for which they bear specific immunoglobulin receptors.

effective, the antigen must be presented in the context of the class II MHC antigens present in the thymus in which the T cell matured.

Cytotoxic T cells are important in immunity against intracellular pathogens such as viruses. These cells recognize antigen in association with MHC class I molecules. Antigen presentation to T_c does not require internalization and digestion. Rather, it appears that productive infection with an intracellularly replicating organism (e.g., a virus) is necessary. Association of viral antigens with class I MHC requires *de novo* protein synthesis in the infected cell. However, the final result is still the association of antigen fragments with MHC (class I, in this case).

T cells are activated by intimate membrane-membrane contact with the APC. The TCR-CD3 complex binds the antigen-MHC complex. In addition, several cell surface proteins serve as "adhesion molecules" enhancing stability of the cellular contact. The CD4 molecule of $T_{h/i}$ cells is a receptor for monomorphic determinants of MHC class II antigens. Similarly, the CD8 molecule of T_c is a receptor for monomorphic class I MHC determinants. The CD2 pan-T cell marker and the CD58 molecule of APCs also appear to have a mutual receptor-ligand relationship. Additional interacting cell-surface molecules may yet be discovered. Whether these interactions serve simply as stabilizers, or are required for cellular activation also remains to be determined.

Thus, the first step in T cell activation in both humoral immune responses (to TD antigens) and cellular responses, is contact with an APC having an appropriate antigen-MHC complex on its surface (Figure 7.7). Subsequent steps in activation require the participation of other cells or factors described below.

SOLUBLE MEDIATORS OF CELLULAR COOPERATION: THE CYTOKINES

Cytokines are protein or glycoprotein factors secreted by cells of both monocytoid (monokines) and lymphoid (lymphokines) lineages. These factors act on leukocytes (and many other cell types) bearing the corresponding receptors, and influence cell division, differentiation, metabolism, motility, and expression of immune effector functions. Several cytokines have been defined to date; no doubt, many more will be discovered in the near future. The immunological effects of some cytokines are summarized in Table 7.II.

Table 7.II. CYTOKINE ACTIONS IN THE IMMUNE SYSTEM

Interleukin 1	Activates resting T cells (induces) receptors for IL-2)
	Increases $T_{h/i}$ secretion of IL-2
	Induces production of IL-3, IL-6, IFN-γ and TNFs
	Influences B cell activation
	Activates NK cells
	Activates macrophages
	Increases CSF production in bone marrow
Interleukin 2	Induces proliferation of activated T cells
	Influences B cell activation
	Activates NK cells
Interleukin 3	Promotes granulopoiesis and lymphopoiesis
	Mast cell growth factor
Interleukin 4	Increases MHC class II on resting B cells
	Influences B cell differentiation
	Influences activation of mature B cells
	Mitogen for $T_{h/i}$ and T_c cells
	Induces antigen-specific T_c
	Synergizes with IL-3
Interleukin 5	Influences B cell differentiation
	Stimulates eosinophil differentiation
Interleukin 6	Influences B cell differentiation
	Increases IgM, IgG, IgA secretion by antibody-secreting cells
	Induces T_c differentiation
	Influences hemopoiesis
	Induces acute phase reaction
Interleukin 7	Induces proliferation of immature B cells and T cells
	Induces proliferation of mature T cells
Interleukin 8	Chemotactic for PMNs and T cells
	Induces PMN degranulation
Interferons, α, $\beta 1$	Increases MHC class I expression
	Activate NK cells
Interferon γ	Increases MHC class I and II expression
	Activates macrophages, NK cells, and T_c
	Increases IgG FcR expression
Tumor necrosis factors α, β	Synergize with IL-1 systemic effects
	Cytotoxic factor?

Interleukin 1

Interleukin 1 (IL-1) is produced by many different cell types. These include monocytes, dendritic cells, Langerhans cells, epithelial cells, endothelial cells, astrocytes, and microglia. IL-1 has a wide range of biological effects. It induces expression of receptors for interleukin 2

(IL-2) on T cells, and stimulates production of IL-2 by $T_{h/i}$ cells; it induces the synthesis of other cytokines (e.g., interleukins 3 and 6, interferon γ, and tumor necrosis factors). IL-1 synergizes with other cytokines in promoting natural killer cell cytotoxicity; it activates macrophages; and it increases production of colony stimulating factors (see Chapter 2) in the bone marrow. In addition to its effects on leukocytes, IL-1 may cause fever (it has also been called *endogenous pyrogen*), and it stimulates margination of neutrophils, release of acute phase proteins by hepatocytes, release of collagenase by chondrocytes, production of plasminogen activator, production of thromboxane in endothelial cells, synthesis of collagen type III in epidermal cells, proliferation of osteoblasts and osteoclasts, and it also has CNS effects such as anorexia or somnolence. Some of these activities attributed to IL-1 may actually be secondarily mediated by IL-6 produced in response to IL-1.

Two forms of IL-1 are produced, IL-1α and IL-1β, encoded by distinct genes located on chromosome 2 in humans. After intracellular processing, a molecule of $M_r = 17{,}000$ is secreted. IL-1 receptors are widespread in many tissues as evidenced by its broad array of biological effects. A single receptor is able to bind either IL-1α or IL-1β. Two types of IL-1 receptor have been identified, one of low affinity and one with high affinity. The biological correlates of the two forms of IL-1 and the two receptors are not known.

Several proteins or glycoproteins have been found to inhibit the activity of IL-1 *in vitro*. The urine of patients with monocytic leukemia contains an IL-1 inhibitor which probably is secreted by the malignant cells. This suggestion is supported by the finding that monocytes activated with immune complexes also produce the inhibitor. Uromodulin, found in the urine of pregnant women, has general immunosuppressive properties; it is another inhibitor of IL-1. Macrophages infected with cytomegalovirus or with human immunodeficiency virus produce large amounts of IL-1, as well as inhibitors of IL-1. Whether or not these inhibitors are part of a feedback regulatory mechanism for IL-1 activity is subject to speculation.

Interleukin 2

Interleukin 2 (IL-2), is a highly glycosylated protein secreted by activated T cells. IL-2 promotes growth of T cells (it is a cytotoxic T cell differentiation factor) and B cells. IL-2 also activates natural killer cells, mature cytotoxic cells, and a population of cells called *lymphokine-activated killer cells* (see below). Human IL-2 has $M_r = 15{,}000$. Its carbohydrate component is not required for activity since recombinant IL-2 which is not glycosylated is active *in vitro*. IL-2 is encoded by a single gene on human chromosome 4.

Resting mature T lymphocytes secrete no IL-2, and contain no IL-2 mRNA. T cell activation by polyclonal mitogens, antigen and APCs, or antibodies reacting with the T cell antigen receptor, leads rapidly to synthesis and secretion of IL-2. All of the above mechanisms of T cell activation depend on IL-1 secreted by accessory cells. IL-1 stimulates IL-2 secretion by $T_{h/i}$ cells, and expression of IL-2 receptors by effector T cells. T cell IL-2 secretion is transient, reaching a peak 6–8 hours after activation, and declining by 12 hours. Most IL-2 is produced by $CD4^+$ T cells ($T_{h/i}$). The IL-2 receptor is a non-covalently linked heterodimer composed of two polypeptides designed α and β.

Activated B cells also express receptors for IL-2. IL-2 increases the production of all Ig isotypes by antigen-stimulated B cells.

Interleukin 3

Interleukin 3 (IL-3) is secreted by activated T cells, and by mast cells following IgE-induced degranulation (see Chapter 10). IL-3 induces differentiation of pluripotent stem cells in the bone marrow by inducing responsiveness to colony stimulating factors (see Chapter 2). IL-3 also increases production of granulocytes and lymphocytes and, in conjunction with IL-4, promotes growth of mast cells. The action of IL-3 is mediated (at least in part) via an increase in histamine production in the bone marrow. IL-3 may also activate mature granulocytes and macrophages. IL-3 is encoded by a single gene on chromosome 5 in a locus containing genes encoding several other cytokines (Figure 7.8). IL-3 receptors are found on bone marrow cells and peripheral blood monocytes.

Figure 7.8. *The cytokine gene complex of chromosome 5.* The genes encoding the cytokines interleukins 3, 4, and 5, and granulocyte-monocyte colony stimulating factor (GM-CSF, see Chapter 2) have been mapped to the long arm of chromosome 5. These four genes are dispersed over a distance of at least one megabase (10^6 DNA base pairs). Chromosomal translocations involving this region have been associated with hemopoietic abnormalities. (Adapted from van Leeuwen et al., 1989.)

Interleukin 4

Interleukin 4 (IL-4) has $M_r = 20{,}000$ and is produced by stimulated T cells, mast cells, and bone marrow stromal cells. IL-4 increases the surface density of MHC class II antigens on resting B cells; it promotes both the development of immature pre-B cells to mature B cells, as well as activation of mature B cells. IL-4 is necessary (but not sufficient) for mature B cell entry into S phase upon activation. It also causes an increase in B cell size and the acquisition of locomotor capacity. IL-4 also has effects on T cells, it induces antigen-specific cytotoxic T cell function, and, like IL-2, can induce the proliferation of $T_{h/i}$ and T_c cells. Additional effects of IL-4 are synergism with IL-3 in promoting growth of mast cells and hemopoietic stem cells, and enhancement of neutrophil phagocytosis and the respiratory burst. IL-4 is encoded by a single gene on chromosome 5 (Figure 7.8). IL-4 receptors have been found on lymphoid and myeloid cells, as well as hepatocytes, and fibroblasts.

Interleukin 5

The major activity of interleukin 5 (IL-5), produced by T_h cells, is to stimulate the differentiation of eosinophils. IL-5 also stimulates B cell proliferation. IL-5 is a protein of $M_r = 21{,}000$, and is encoded by a single gene on chromosome 5 (Figure 7.8). Two types of IL-5 receptor with different affinities have been identified on a B lymphoma cell line.

Interleukin 6

Interleukin 6 (IL-6) is a polypeptide with $M_r = 19–32{,}000$ and has many biological activities. IL-6 induces immunoglobulin secretion by B cells, but not their proliferation; it increases IgM, IgG, and IgA secretion by mitogen-stimulated B cells; it influences the differentiation of hemopoietic stem cells; it induces differentiation of $T_{h/i}$ and T_c cells; it promotes growth of malignant plasma cells (plasmacytomas); and it also has activity similar to nerve growth factor. Its most important physiologic role may be the induction of the acute phase reaction (hepatic synthesis of acute phase proteins and decreased albumin synthesis). Several activities once attributed to IL-1 are now thought to be mediated by IL-6.

IL-6 is produced by T cells, B cells, macrophages, fibroblasts, keratinocytes, endothelial cells, bone marrow stromal cells, mesangial cells, and brain astrocytes. IL-6 is encoded by a single gene located on chromosome 7.

IL-6 increases the rate of cell division of plasmacytomas *in vitro*, and addition to the culture of antibodies binding IL-6 reduces the rate of cell cycling. Abnormal production of IL-6 occurs in some hematologic malignancies, and other disorders. Castleman's disease is a localized or systemic anomalous proliferation of polyclonal plasma cells in lymph nodes or other lymphoid tissue. Overproduction of IL-6 in germinal centers has been implicated in the etiology of this disease. Autoimmune phenomena (see Chapter 10) are frequently associated with a cardiac myxoma (a type of connective tissue tumor); some myxomas produce IL-6. The synovial fluid of patients with rheumatoid arthritis contains elevated levels of IL-6. Both synovial cells and infiltrating T cells are stained with labelled antibodies binding IL-6. Kidney mesangial cells may also produce IL-6, and patients with mesangial poliferative glomerulonephritis have increased amounts of IL-6 in their urine. Renal adenocarcinoma (also called hypernephroma or Grawitz tumor) has also been shown to secrete IL-6.

IL-6 receptors are distributed widely in the body. They are found on hemopoietic precursors, resting or activated B cells, resting T cells, myeloid cells, and hepatocytes. Receptors will probably be found in other tissues as well. As with several other cytokines, there appear to be high-affinity and low-affinity receptors.

Interleukin 7

Interleukin 7 (IL-7) is a 25 kd protein produced by bone marrow, thymus, and spleen stromal cells. IL-7 delivers a purely proliferative signal to early B cells. It also induces proliferation of immature thymocytes (double-negative cells), resting T cells, and some mature $CD4^+$ and $CD8^+$ clones. The gene encoding IL-7 and the IL-7 receptor have yet to be characterized.

Interleukin 8

Interleukin 8 (IL-8) is a small (6.5 kd) protein which is chemotactic for neutrophils and T cells. IL-8 also induces release of lactoferrin from neutrophil specific granules.

Interferons

These molecules were initially described as proteins able to "interfere" with viral entry and replication. Interferons are a group of low molecular weight proteins produced by cells during infection with viruses, rickettsiae, mycoplasma, protozoa, and fungi, and in response to a variety of exogenous stimuli such as bacterial endotoxins, nucleic acids or polyanions such as dextran. Interferons have several effects on the functions of immunocompetent cells, in addition to their actions inducing resistance to many viral infections in most tissues.

Interferons (IFNs) are divided into two types, type 1 and type 2. Type 1 IFN consists of α interferon and $\beta 1$ interferon (what previously was known as interferon $\beta 2$ is now called IL-6). Type 2 interferon is usually referred to as γ interferon. IFN-α and IFN-$\beta 1$ bind to the same receptor, while IFN-γ has a distinct receptor. Many cell types throughout the body have IFN receptors, and can themselves synthesize interferons when infected by certain viruses. Some cell types may produce and secrete relatively large quantities of IFNs. IFN-γ has several properties distinct from the type I IFNs, and plays a significant role in antibody responses.

Type 1 interferon (α and $\beta 1$)

The name interferon α (IFN-α) is given to a group of about 20 related proteins, also known as *leukocyte interferon* since these cells produce it in relatively large quantities. IFN-$\beta 1$ has also been called *fibroblast interferon* for similar reasons. These IFNs increase expression of class I MHC antigens on cell surfaces, and stimulate natural killer cell activity. Type 1 IFNs generally inhibit cell division, and may be immunosuppressive in large quantities. The α and $\beta 1$ interferons are encoded by distinct genes located on chromosome 9.

Type 2 interferon (γ)

Also known as *immune interferon*, IFN-γ has several effects. It increases expression of class I and II MHC antigens, and stimulates macrophage activity. It also increases expression of receptors for IgG Fc ($Fc_{\gamma}R$), activates natural killer cells and cytotoxic T cells. In high doses, IFN-γ is immunosuppressive, analogous to the type 1 IFNs.

Several years ago, γ interferon was heralded by the media as a potential new miracle therapy for cancer. This enthusiasm was based on interferon's activation of natural killer cells, and the belief that these cells were central in host defences against tumors. The enthusiasm appears to have been premature, however. The results of several clinical trials were inconclusive, at best.

Recently, IFN-γ has been shown to be effective in treating chronic granulomatous disease (CGD) type IA (see Chapter 4). IFN-γ induced the synthesis of cytochrome b heavy chain, augmented the respiratory burst, and increased production of superoxide anions in neutrophils and macrophages *in vitro*. Subcutaneous injection of IFN-γ produced the same improvements in phagocyte function in patients. IFN-γ is currently being investigated further for its efficacy in reducing the severity of CGD.

Antiviral action of interferons

Interferon induces the production of two enzymes which may affect synthesis of viral proteins during infection. These are a protein kinase, and 2′–5′ oligoadenylate synthetase. Double-stranded RNA (dsRNA) appears to be required for activity of these proteins. While dsRNA does not normally occur in transcription and translation in eukaryotic cells, it is found as an intermediate in replication of many viruses. The protein kinase activities eukaryotic initiation factor-2 (EIF-2), thereby inhibiting protein synthesis. The 2′–5′ oligo-A synthetase forms 2′–5′ oligoadenylic acid, a cofactor needed for activity of an endogenous ribonuclease, RNAse L. RNAse L degrades messenger and ribosomal RNAs.

IFN-α has found clinical application in the therapy of hairy cell leukemia, and chronic viral hepatitis.

Tumor necrosis factors

There are two types of tumor necrosis factor (TNF), TNFα (also called *cachectin*) and TNFβ (also called *lymphotoxin*). These molecules have $M_r = 17{,}000$. Although these proteins are only 28% homologous, they bind to the same receptors and have similar effects. TNFs share many of the biological activities of IL-1, and appear to synergize with it. Macrophages secrete TNFα, activated T_h cells secrete TNFβ. Bacterial endotoxin stimulates macrophages to produce more TNF mRNA, but no protein is made. A second signal delivered by IFN-γ and/or IL-1 or IL-6 is required for TNF synthesis.

Some of the harmful systemic effects of large doses of TNFs are fever, diarrhea, hypotension, metabolic acidosis, hemoconcentration, and hemorrhagic necrosis of several organs, especially the lungs. Sufficiently large doses lead to shock and death. At least some of these effects are due to widespread activation of neutrophil cytotoxic activity. In animal experiments, when chronically administered in very small doses, TNFs cause cachexia, i.e., anorexia, anemia, and rapid weight loss.

Some have proposed that TNFs are important in responses to infectious agents and in tumor immunity. TNFα ameliorates murine malaria infections and potentiates neutrophil fungicidal activity *in vitro*. Abnormally high levels of TNFα have been found in patients with mucocutaneous lymph node syndrome (Kawasaki disease). However, it is not clear if this is a cause of pathology, or a product of another process responsible for tissue injury.

Numerous additional cytokines have been recently described, more and more are being discovered every day. As can be seen from several examples above, there are important interconnections between the immune system and various other tissues mediated by cytokines. These interconnections become more complex with each new factor discovered. Unraveling these intricate cytokine-determined biological relationships will be an extremely challenging endeavor in the future.

Although their biology is still quite incompletely understood, many of the above-mentioned cytokines are being used clinically in attempts to favorably alter the course of malignancy, immunodeficiency or autoimmunity. Cytokines are a subset of what have come to be called *immunomodulators*, that is, substances capable of influencing immune system function. Monoclonal antibodies directed against surface molecules of pathological (e.g., malignant) cells, or cells (e.g., lymphocytes) involved in the immune responses against them are another type of immunomodulator.

Cytokines and T cell activation

Interleukins 1, 2, 4, 6, and 7 influence proliferation and differentiation of precursors of (or mature) MHC class II-restricted T cells ($T_{h/i}$). Interleukins 2, 4, 6, 7, and IFN-γ have been shown to influence the proliferation and differentiation of precursors of (or mature) T_c cells (class I-restricted). While IL-1 induces appearance of receptors for IL-2 on $T_{h/i}$ cells, it does not have this effect on T_c. Another molecule called IL-2 receptor inducing factor (RIF) has this function. RIF is distinct from IL-1, 2, 3, 4, and IFN-γ.

A phosphatidylinositol pathway identical to that operating in B cells is important in early events in T cell activation. Elucidation of additional biochemical mechanisms in T cell activation awaits further research.

CELL-MEDIATED IMMUNITY

Cytotoxicity

Cytotoxic T lymphocytes

Resting cytotoxic T lymphocytes (CTLs or T_c) are functionally inert. Once activated by antigens (associated with APC MHC) or lectins, these cells acquire the ability to lyse target cells bearing antigen on their surface. CTLs express the CD8 marker. All of the cytologic changes from resting CTL to activated CTL are not known, but probably involve some blast transformation, proliferation, and the loss and acquisition of particular cell-surface proteins. In addition to the above-mentioned activating stimuli, CTLs may also be activated by antibodies against idiotypic or allotypic determinants of the T cell receptor, or by antibodies against molecules associated with the TCR such as CD3.

CTLs recognize antigen in the context of self MHC class I gene products. Thus, CTLs lyse autologous cells infected by an intracellularly replicating pathogen such as a virus, and which have viral proteins associated with cell surface class I molecules (Figure 7.9). *Allospecific* or *alloreactive* CTLs may also lyse allogeneic cells directly (without additional antigen). Presumably, allo-MHC class I molecules appear to CTLs as would self class I complexed with a foreign peptide. CD8 molecules bind to a monomorphic determinant of class I antigens and enhance the contact between T_c and their targets. Some cytotoxic T cells express CD4, rather than CD8. This is a subset of alloreactive CTL which lyse target cells expressing foreign MHC class II molecules. We will have more to say about these cells in Chapter 8.

Some question remains whether or not T_c cells require intimate contact with $CD4^+$ cells following antigen presentation in order to express their effector function. Some argue that this is not the case since

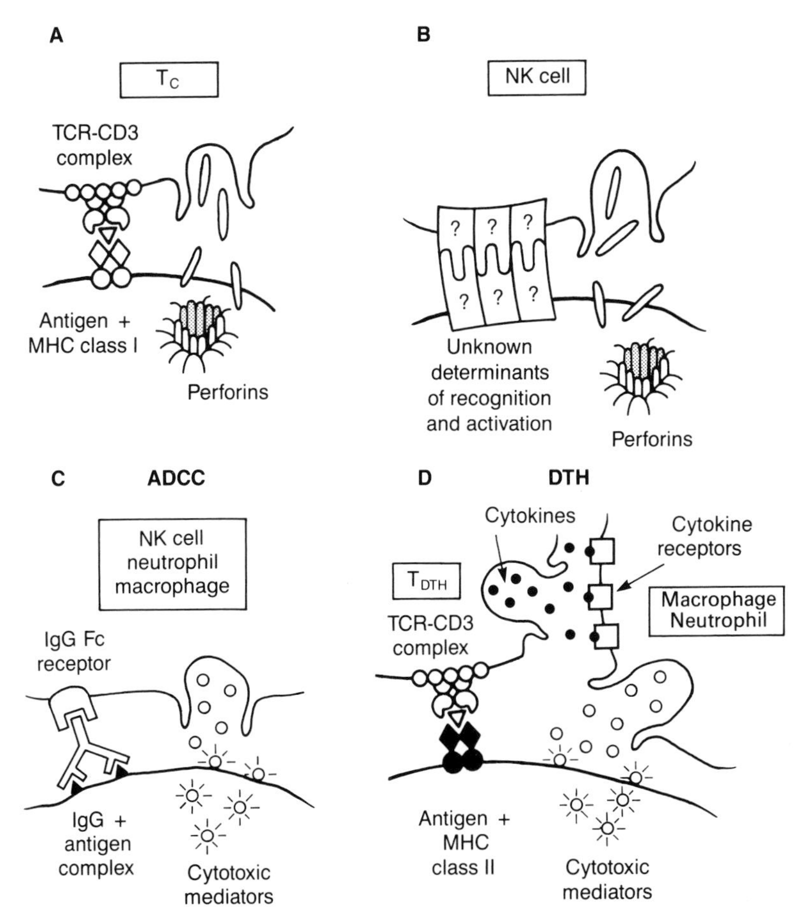

Figure 7.9. *Mechanisms of cytotoxicity.*

A. *CTL-mediated lysis.* The CTL's TCR-CD3 complex binds the MHC class I-peptide complex on the surface of the target cell. The intercellular contact also involves adhesion molecules such as CD8 (the CTL molecule interacting with monomorphic MHC class I determinants). Cytolysis is mediated by perforins (and proteases) stored in CTL granules.

B. *NK cell-mediated lysis.* NK cells use a cytolytic mechanism very similar to CTLs (perforins). However, the surface characteristics of target cells which determine NK recognition and activation are not yet known.

C. *Antibody-dependent cellular cytotoxicity (ADCC).* NK cells, neutrophils and macrophages bear receptors for IgG FC. IgG bound to the target cell's surface becomes a bridge between the target and lytic cells (much as in opsonization, see Figure 4.5). Neutrophils and macrophages do not contain perforins and lyse cells with proteases and other cytotoxic molecules yet to be characterized. Whether or not NK cells use perforins in ADCC remains to be seen.

D. *Delayed-type hypersensitivity (DTH).* $CD4^+$ T_{DTH} recognize antigen in association with MHC class II molecules. Activated T_{DTH} release a multitude of cytokines which recruit and activate mononuclear cells (monocytes and macrophages) as well as neutrophils which release vasoactive and cytotoxic molecules leading to edema and tissue destruction.

resting cytotoxic precursors do not express MHC class II molecules (they are acquired after activation). A recently proposed model suggests that both T_i and T_c cells are activated simultaneously by a single APC (a ternary cellular complex). Although the two T cells do not actually make membrane-membrane contact, they are held in close proximity, and the T_i secretes cytokines inducing maturation of the precursor to a cytotoxic effector cell.

The first step in lysis is the intimate apposition of the plasma membranes of the cytotoxic and target cells. The Golgi apparatus of the cytotoxic cell reorganizes and polarizes secretion toward the target cell, granules are exocytosed into a narrow space between the cell membranes. Cytotoxic cell granules contain *perforins.* These proteins insert into the target cell membrane, polymerize and form channels. These pores are very similar to the membrane lesions produced by complement proteins; perforins and C9 have structural similarities.

Tumor necrosis factors have been implicated in additional mechanisms of membrane lysis. Other granule proteins which may be important for lysis are serine esterases called *granzymes* (granule enzymes) and phospholipase A_2. The latter may act by degrading membrane phosphatidylcholine to lysolethicin, a powerful detergent.

A single cytotoxic cell may "recycle" and lyse tens of target cells without itself suffering damage. Since cytotoxic cells secrete their granule contents into a space between themselves and their targets, how are they protected from the destructive potential of the molecules they discharge? The answer to this question remains somewhat obscure. Cytotoxic cells are not themselves constitutively immune to lysis since they may be the targets of other cytotoxic cells. A protein known as *homologous restriction factor* (*HRF*) protects cells (e.g., erythrocytes) from lysis by the membrane attack unit of complement. An identical or highly similar protein which inhibits perforin-mediated lysis has been isolated from cytotoxic T cell granules. This molecule may protect the cytotoxic cell from its own perforins.

Memory in the cellular response

As are humoral immune responses, cellular responses are generated more rapidly and are of greater magnitude during the second and subsequent exposures to antigen. Recently, several cell-surface proteins have been investigated as markers for memory *versus* virgin T cells.

The cell surface protein CD45 is a family of related polypeptides present on all leukocytes (it has been called *leukocyte common antigen*). However, forms of differing molecular weights serve to distinguish different leukocyte populations. A form with $M_r = 220$ kd is CD45RA, a marker for B cells and some T cells (as well as granulocytes and macrophages). Another form with $M_r = 180$ kd is called CD45RO. Most T cells from the blood of newborns express large amounts of CD45RA (CD45RA-high), and have very little CD45RO. On the other hand, the expression of small amounts of CD45RA (CD45RA-low) is associated with expression of large amounts of CD45RO and other markers associated with T cell activation (such as the adhesion molecules CD2, CD58, and CD11a/CD18). For these reasons, CD45RA-high cells are postulated to be naive or virgin T cells, and CD45RA-low to be memory T cells.

Natural killer (NK) cells

These cytotoxic cells derive their name from their discovery in mice that had not been intentionally immunized, hence, they were "naturally occurring" cytotoxic cells. These bone marrow-derived cells are described morphologically as large granular lymphocytes (LGL), and have not been shown to possess unique cytodifferentiation antigens. CD8, CD11, CD16 (FC_γRIII), CD56, and CD57 have been found on NK cells, but some of these antigens may be found on granulocytes, monocytes and activated T cells. Because of their cytotoxic activity, however, many researchers would classify NK cells as a type of T cell.

There is much evidence that NK cells lyse target cells by the same molecular mechanisms as T_c cells (see above), although they clearly differ in their requirements for activation (Figure 7.9). NK cells do not express TCR-type antigen receptors, however they do bear on their surfaces the ζ (zeta) chain component of the CD3 complex, suggesting that they do have some functionally equivalent structure. Furthermore, NK-target recognition does *not* depend on MHC molecules. The precise determinants of NK recognition are unknown. Recent studies have suggested that a component of MHC class I molecules called *β_2-microglobulin* may be at least one determinant of NK-target recognition.

NK cells are able to lyse a variety of cells regardless of species origin, thus, they appear to lack any classical immunologic "specificity". One of the more intriguing aspects of their function is the preferential lysis of neoplastic cells. NK cells lyse some untransformed targets, but in general they lyse tumor cells more rapidly and at lower effector:target ratios. This characteristic has led to speculation that NK cells are important in protection from neoplasms. This speculation has yet to be firmly upheld or refuted. NK cells also appear to be important in the control of certain virus infections. An adolescent patient with extreme susceptibility to disseminated herpesvirus (varicella-zoster, HVZ) infections was found to be completely lacking in NK cells. However, she had normal levels of specific humoral and cellular immunity to HVZ. NK function is stimulated by interferons and IL-2, and may be influenced by regulatory T cells.

T cells bearing TCR1 (γ/δ), and having the phenotype $CD3^+/CD4^-/CD8^-$ (double-negative) and $CD56^+$ have been shown to have two cytotoxic mechanisms. One of these corresponds to the "classical" specific T cell type of cytotoxicity, and the other to the NK mode of cytotoxicity. These cells may be a functional "cross" between NK cells and T cells. Some researchers have suggested that NK cells are evolutionary precursors of clonally distributed "specific" cytotoxic T cells.

Lymphokine-activated killer cells (LAK)

When cells from virtually any lymphoid tissue are exposed to high concentrations of IL-2, there appears a population of cells having the ability to destroy neoplastic cells. These are called lymphokine-activated killer cells (LAK). A wide variety of cells may become LAK. Under the influence of IL-2, T and B lymphocytes and NK cells all have the ability to become cytotoxic for tumor cells. The molecular events leading to LAK activity, details of the cytotoxic mechanism, as well as the determinants permitting recognition of neoplastic *versus* normal cells all remain unknown, at present. LAK are currently being tested in several clinical trials for efficacy in the treatment of various cancers.

Antibody-dependent cell-mediated cytotoxicity (ADCC)

As its name implies, this immune mechanism is a cytocidal reaction dependent on the presence of target cell-bound antibodies. These antibodies are believed to link the cytotoxic cell and target together by simultaneous binding of a target cell surface antigen, and the F_c receptor of the effector cell (just as in opsonization, see Figure 4.5). This function may be mediated by neutrophils, macrophages, and NK cells (Figure 7.9).

Since ADCC may be mediated by several different cell types, it may also occur via several different mechanisms. Perforin-like molecules have not been found in macrophages or PMNs, however, PMNs do form membrane-membrane contacts with their targets. Lysis occurs at least in part by release of proteolytic enzymes. Whether NK cells use perforin-dependent mechanisms in ADCC remains to be seen. ADCC is stimulated by interferons and IL-2.

The physiological role of this reaction is not well-established. Some speculate that ADCC may also operate in tumor immunity. This cytotoxic reaction is quite evident in graft rejection, and in a number of autoimmune diseases as well (see Chapter 10).

Delayed type hypersensitivity

The delayed type hypersensitivity reaction was originally observed in the tuberculin skin reaction developing in guinea pigs sensitized with tubercle bacilli (*Mycobacterium tuberculosis*). The reaction is no different in humans injected intradermally with the *purified protein derivative* (*PPD*) of *M. tuberculosis* after sensitization by natural infection. Approximately 48 hours after injection (hence, "delayed"), the site becomes erythematous and indurated. Histological study reveals accumulation of mononuclear cells (80–90% monocytes) around small blood vessels. Although rare in routine tuberculin skin testing, severe hypersensitivity reactions can lead to tissue necrosis.

DTH reactions may be transferred from sensitized to unsensitized animals with $CD4^+$ T cells (which recognize antigen in association with MHC class II molecules). T_{DTH} cells react with antigen-presenting cells at the site of injection, proliferate and release a multitude of cytokines (Figure 7.9).

Some of these factors are chemotactic, each of the different types of leukocyte having its own specific chemoattractant (e.g., *neutrophil chemotactic factor*, *basophil chemotactic factor*, etc.). Other factors increase small vessel permeability facilitating leukocyte diapedesis (e.g., *lymph node permeability factor*). Yet other factors cause incoming leukocytes to cease movement (*migration inhibition factor*). Interferons stimulate leukocytes to synthesize and secrete cytotoxic substances and inflammatory mediators (see Chapter 10), as well as release those performed and stored in granules.

DTH may be either a local or systemic reaction. The local reaction is exemplified by the tuberculin skin test described above. Systemic DTH occurs when a large amount of antigen enters the blood, as in septicemia. Systemic DTH is manifested by fever, malaise, myalgia, arthralgia, and lymphopenia. Severe cases may lead to shock or death.

Contact sensitivity

The contact sensitivity reaction (CSR) is a form of delayed hypersensitivity in which the targets are cellular elements of the skin. Substances inducing CSR have two major characteristics: low molecular weight, allowing them to diffuse into the skin, and the ability to combine chemically with skin proteins. A classic example is the reaction to *urushiol* from certain plants of the genus *Toxicodendron* (a.k.a. poison ivy). Man-made chemicals such as picrylchloride and dinitrochlorobenzene may also induce CSR. These molecules combine with skin proteins and change their structure so that they are no longer regarded as self components. In the normal process of epidermal cell death and keratinization, these proteins are taken up by skin macrophages and presented to T cells. Thus, T cells are activated and release cytokines recruiting and activating the effector cells, the monocytes and macrophages.

CSR in its mildest form is simply erythema and induration. More severe reactions lead to vesiculation and necrosis. The major difference between CSR and DTH is that in CSR, the entire process is limited to the skin, since that is the only location where the antigen is found. T cells mediating contact sensitivity are frequently called T_{CS}. Whether these are a subset of cells distinct from T_{DTH} remains to be determined.

A variant of CSR is *cutaneous basophil hypersensitivity* (*CBH*) also known as *Mote-Jones hypersensitivity*. This reaction is characterized by a basophil infiltration occurring after intradermal antigen injection. This is a much milder reaction with intense pruritus and slight skin edema, and it usually abates in 24–48 hours. CBH may be accelerated and itensified by application of heat to the affected area. In marked

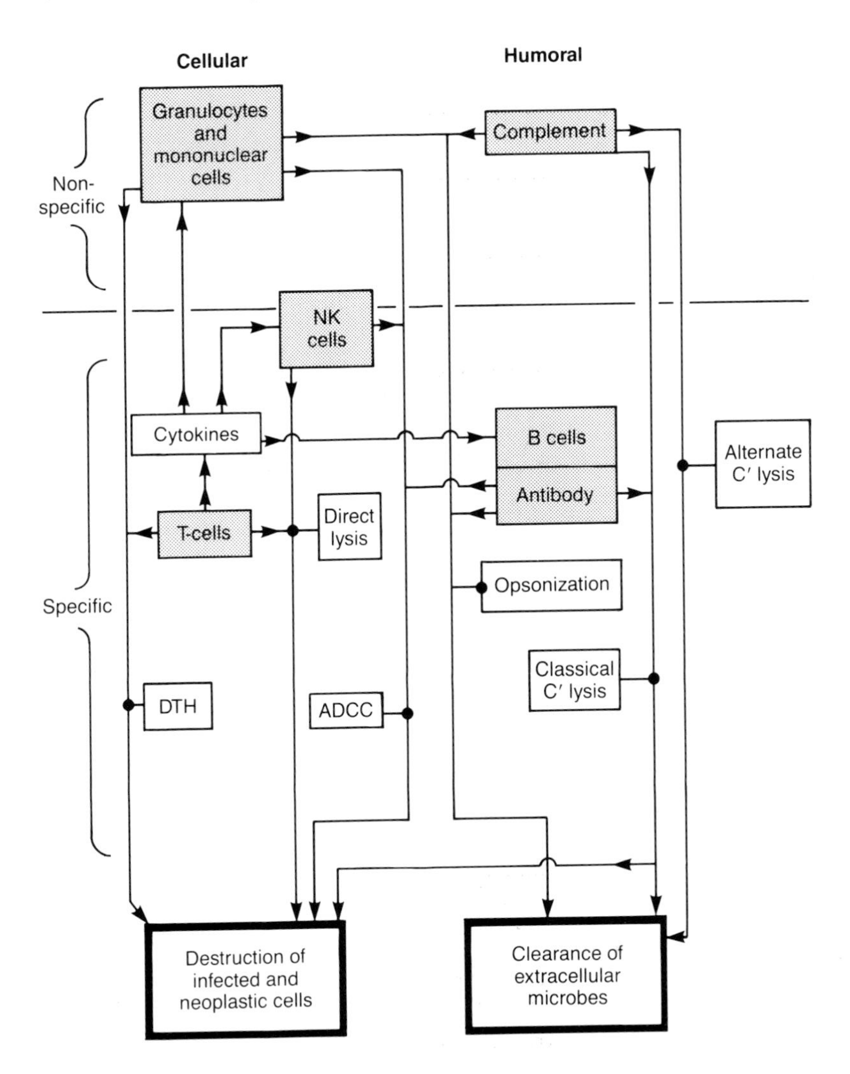

Figure 7.10. *Interrelationships between non-specific and specific humoral and cellular immunity.* Some mechanisms of non-specific immunity act independently of specific immunity. The alternate pathway of complement and the phagocytic activity of granulocytes, monocytes/macrophages and other cells of the mononuclear phagocyte system may destroy microbes without help from the products of specific immunity. However, several specific and non-specific systems interact. Specific antibody initiates the classical pathway of complement and stimulates phagocytosis by opsonization, and initiates ADCC. Through the actions of cytokines, various T cell populations also influence the activity of cellular components involved in all phases of immunity. NK cells appear to have some characteristics of both non-specific and specific mechanisms.

contrast to DTH and the CSR, CBH is dependent on serum antibody, and shows no MHC restriction.

Four mechanisms of cytotoxicity are summarized in Figure 7.9: CTL- (T_c-) mediated lysis, NK cell-mediated lysis, ADCC, and DTH. Now that we have introduced all of the basic elements of humoral and cellular immunity, Figure 7.10 illustrates some of the interrelationships between these protective mechanisms.

CELLULAR REGULATION OF IMMUNE RESPONSES

Humoral and cellular immune responses result from antigen-triggered clonal proliferation. If antigen stimulation were to persist (as in some infections), might not the amplifying mechanisms we have described cause expanding clones to overwhelm the body? Clearly, this does not happen during a physiological immune response. Only in instances of aberrant regulation (autoimmune disease) or malignant transformation (lymphoma, leukemia) do we become victims of over-exuberant clonal proliferation.

The extinction of an immune response is *not* simply a passive phenomenon of clearance of antigen from the blood and lymph and its degradation by phagocytic cells. Much research indicates that a special type of regulatory T cell exerts a negative feedback on both humoral and cellular immunity. This negative regulatory T cell is called a *suppressor T cell* (*T_s*). T_s precursors can be found in the thymus as well as in peripheral lymphoid tissue.

T_s cells are the focus of some controversy in immunology. Some researchers dispute their very existence since a great many facts concerning the manner in which they interact with one another and with more well-defined lymphocyte populations remain mysterious despite intense investigation. These calls do not use the receptor genes used by $T_{h/i}$ and T_c, and many aspects of their function do not appear to follow the patterns of other types of T cells. We here mention briefly a few concepts culled from a large and complex literature.

Three different types of T_s cells constituting a *suppressor cell circuit* have been defined by studies of antigen-specific suppression. The first is an *inducer of suppression*, also called T_s1. The second is a *transducer cell*, or T_s2. The third is the *suppressor-effector cell*, or T_s3. The T_s1, being an inducer type of T cell, expresses CD4. Some have distinguished human suppressor-inducers from helper/inducers on the basis of three cytodifferentiation antigens: Leu8, CDw29, and 2H4. Helper/inducers are $Leu8^-/CDw29^+/2H4^-$ while suppressor-inducers are $Leu8^+/CDw29^-/2H4^+$. When speaking of suppressor cells (T_s), one is taken to mean effectors of suppression (T_s3) unless one specifies otherwise. The T_s3, like T_c, express CD8. Human T_s3 may be distinguished from T_c in that the former are $CD11^+/Tp44^-$ while the latter are $CD11^-/Tp44^+$. The T_s2 has not been well-characterized.

T_s1 are presumably activated by antigen presenting cells as are $T_{h/i}$. An important difference between their activation requirements is that inducers of suppression appear not to be MHC-restricted. Activated T_s1 then secrete several soluble factors, including cytokines such as *T suppressor cell growth factor*, IL-2, and IFN-γ, activating T_s2 and/or T_s3. T_s3 may exert their effects either on $T_{h/i}$, directly on B cells, or on effector T cells such as those mediating DTH.

Different participants may be implicated in suppressor circuits depending upon the system under study. One may distinguish two, three, or more types of regulatory cells in a particular circuit. In addition, two other types of regulatory cells have been suggested by the phenomenology of suppression. These are the *contrasuppressor T cell* and the *veto cell*. Contrasuppressor cells are $CD8^+$ T cells which "liberate" $CD4^+$ cells from suppression by effector T_s. Veto cells have been described as a population of cells responsible for maintaining self-tolerance (see Chapter 9).

Suppressor circuits may be categorized according to the reactions they suppress, or the specificity of the effectors. Thus, we distinguish nonspecific, antigen-specific, and immunoglobulin-specific subsets. Non-specific T_s, as their name implies, may suppress immune responses irrespective of antigen specificity. An excess activity of non-specific T_s has been suggested as one mechanism in the etiology of *common variable hypogammaglobulinemia*. Antigen-specific T_s, on the other hand, suppress only the response to a particular antigen. Immunoglobulin-specific T_s may suppress the expansion of B cell clones expressing a particular isotypic, allotypic, or idiotypic determinant in an immune response. Both immunoglobulin and TCR idiotypic determinants may be recognized by T_s. One of the contributing factors to some autoimmune diseases (see Chapter 10) such as sytemic lupus erythematosus may be an abnormally low activity of T_s cells.

With some good fortune, perhaps, the convoluted phenomenology of suppressor T cells will soon more clearly reveal itself to researchers.

DEFECTS OF CELLULAR IMMUNITY

Congenital T cell immunodeficiencies

Congenital thymic aplasia

T cell development requires a functional thymus, derived during embryogenesis from the third and fourth pharyngeal pouches. These structures also give rise to the parathyroid glands. Congenital thymic aplasia, or the *DiGeorge syndrome*, results from failure of this step in embryogenesis. A diagnosis of DiGeorge syndrome is not first suspected because of immunodeficiency, but because of profound hypocalcemia due to an absence of the parathyroid glands and parathormone. Hypocalcemic tetany is not uncommon in the first days of life. The number and function of T cells are drastically reduced. Afflicted infants are susceptible to infections by intracellular pathogens such as viruses. This disease can be corrected only with transplants of fetal thymic tissue.

Purine nucleoside phosphorylase deficiency

Patients with this congenital metabolic deficiency have a low number of T cells, and normal numbers of B cells, and antibody production is generally adequate. These children have severe susceptibility to viral infections, fatalities have been observed with varicella virus. The absence of this phosphorylase causes an accumulation of deoxyguanosine triphosphate which, in turn, inhibits the activity of ribonucleotide reductase. These metabolic changes lead to a selective inhibition of T cell division.

Wiskott-Aldrich syndrome (WAS)

Patients with this rare X-linked disease have an immune deficiency associated with thrombocytopenia and eczema. The most common immune abnormality in patients is lack of antibody response to polysaccharide antigens (e.g., pneumococcal capsular polysaccharides). However, humoral responses to most protein antigens appear to be intact. There is often a characteristic anomalous pattern of isotype production with very elevated IgA and IgE, normal or somewhat high IgG, and low IgM. There is also frequently an impaired DTH response to various antigens (e.g., dinitrofluorobenzene), as well as decreased T cell response to mitogens. The molecular defect(s) responsible for this disease are not known. However, the restoration of anti-polysaccharide antibody production with T cells from a normal individual in conjunction with WAS B cells suggests a primary T cell defect.

WAS usually first manifests itself as a bleeding disorder at about six months of age. Despite antibody responses to many pathogens, WAS patients are susceptible to severe infections with cytomegalovirus, varicella, and herpes simplex. Bacterial otitis and pneumonia are also common.

Platelet transfusions may help to control bleeding. Intravenous immunoglobulin and antibiotic therapy help prevent and treat infection. These measures are not very effective, however, and with only these treatments survival is usually to about 3–4 years of age. Bone marrow transplantation with a suitably matched donor provides the greatest potential for prolonging life.

The acquired immunodeficiency syndrome (AIDS)

The most prevalent T cell immunodeficiency today is AIDS. This syndrome was first described in 1981. The characteristic susceptibility to *opportunistic infections* (caused by organisms commensal or not pathogenic in healthy individuals) and the relatively high incidence of certain neoplasms (e.g., Kaposi's sarcoma) was first observed in intravenous drug abusers and homosexual males. This syndrome results from infection of $CD4^+$ (helper/inducer) T cells by a retrovirus. Viral replication destroys these cells which have a crucial function in generating immune responses to most infectious agents. The etiologic agent of AIDS is a retrovirus known as *human immunodeficiency virus* (*HIV*). There are two types, *HIV I* and *HIV II*. The virus receptor is the CD4 protein on the $T_{h/i}$ cell surface.

It was recently discovered that the AIDS virus may infect monocytes or macrophages as well as $T_{h/i}$ cells. The extent to which HIV replicates, and whether or not it is cytopathic within monocytes is not known. Infection of monocyte-derived cells may be a reservoir within the body for the AIDS virus.

Epidemiologic data indicate that the groups of people most at risk for AIDS infection are intravenous drug abusers, promiscuous homosexual males, hemophiliacs receiving factor VIII concentrates, and babies of infected women. The transmission of AIDS by heterosexual contacts is well-documented, though its frequency is much debated. Transmission from men to women appears to be more frequent than from women to men. The virus has been isolated from blood, semen, saliva, tears, breast milk and urine. The incubation period between detection of the virus in lymphocytes or the appearance of anti-viral antibodies, and exhibition of the symptomatology of AIDS is highly variable. This period may range from a few months to as many as 15 years.

The principal phenomenon underlying the pathology of AIDS is the gradual depletion of cells bearing the CD4 surface antigen: helper/inducer T cells. Subclinical immunodeficiency is manifested by non-specific symptoms such as fever, weight loss, diarrhea, fatigue, and skin lesions. More striking symptoms arise after infection by a pathogen or a commensal organism (opportunistic infection), or appearance of a neoplasm. The major opportunistic infections observed in AIDS are listed in Table 7.III. Neurological symptoms may also occur in AIDS due to infection of glial cells bearing the CD4 antigen. The function of the CD4 molecule on these cells is unknown.

Usually the ratio of $CD4^+$ to $CD8^+$ cells in peripheral blood is greater than one. Inversion of the CD4/CD8 ratio is characteristic of HIV infection. Other laboratory indications of infection include low or absent proliferative responses to T cell mitogens; low responses to T-dependent antigens; presence of serum antibodies to HIV surface glycoproteins; appearance of reverse transcriptase activity in lymphocytes; and presence of viral nucleic acid in lymphocyte DNA.

There is yet no effective virus-specific therapy for HIV or any other retrovirus infection. Much AIDS-related research is focused on

Table 7.III. OPPORTUNISTIC INFECTIONS IN AIDS

Pneumocystis carinii pneumonia
Cerebral toxoplasmosis
Intestinal cryptosporidiosis, isosporosis
Candidal esophagitis, thrush
Cryptococcal meningitis
Disseminated infections with:
cytomegalovirus
herpes simplex viruses
herpes varicella-zoster virus
Epstein-Barr virus
histoplasma, coccidioides
M. tuberculosis
atypical mycobacteria

(Adapted from Kovacs and Masur, 1988.)

developing an effective vaccine. Although HIV viruses present daunting obstacles to control by modern biotechnology, the rate at which information about the structure and replication of these viruses has accumulated is unparalleled in the history of science. The pace of AIDS research shows little sign of slowing. Doubtless, even if effective therapy or vaccines for AIDS are not forthcoming in the near future, a wealth of knowledge of the molecular biology and physiology of the immune system will be gained.

We next turn our attention to the system of histocompatibility glycoproteins. As has been alluded to several times in this and preceding chapters, this group of cell surface molecules plays a role in almost every aspect of both humoral and cell-mediated immunity.

SOURCES AND SUGGESTED ADDITIONAL READING

Alexander, G. (1989) Treatment of acute and chronic hepatitis. *Baillieres Clin. Gastroenterol.*, **3**:1–20.

Amman, A. J. & Hong, R. (1989) Disorders of the T-cell system. In E. R. Stiehm, ed., *Immunologic Disorders in Infants and Children*, Third Edition, W. B. Saunders Company, Philadelphia, pp. 257–315.

Berzofsky, J. A., Brett, S. J., Streicher, H. Z. & Tajkahashi, H. (1988) Antigen processing for presentation to T lymphocytes: function, mechanisms, and implications for the T cell repertoire. *Immunol. Rev.*, **106**:5–31.

Biddison, W. E. L Shaw, S. (1989) CD4 expression and function in HLA class II-specific T cells. *Immunol. Rev.*, **109**:5–15.

Bierer, B. E., Greenstein, J. L., Sleckman, B., Ratnofsky, S., Peterson, A., Seed, B. & Burakoff, S. J. (1988) Functional analysis of CD2, CD4, and CD8 in T-cell activation. *Ann. N. Y. Acad. Sci.*, **532**:199–206.

Biron, C. A., Byron, K. S. & Sullivan, J. L. (1989) Severe herpesvirus infections in an adolescent without natural killer cells. *N. Engl. J. Med.*, **320**:1731–1735.

Boey, H., Rosenbaum, R., Castracane, J. & Borish, L. (1989) Interleukin-4 is a neutrophil activator. *J. Allergy Clin. Immunol.*, **83**:978–984.

Cassell, D. & Forman, J. (1988) Linked recognition of helper and cytotoxic antigenic determinants for the generation of cytotoxic T lymphocytes. *Ann. N. Y. Acad. Sci.*, **532**:51–60.

Cerami, A. & Beutler, B. (1988) The role of cachectin/TNF in endotoxic shock and cachexia. *Immunol. Today*, **9**:28–31.

Clevers, H., Alarcon, B., Wileman, T. & Terhorst, C. (1988) The T cell receptor/CD3 complex: a dynamic protein ensemble. *Ann. Rev. Immunol.*, **6**:629–662.

Dinarello, C. A. (1988) Interleukin-1. *Ann. N. Y. Acad. Sci.*, **546**:122–132.

Dinarello, C. A., Clark, B. D., Puren, A. J., Savage, N. & Rosoff, P. M. (1989) The interleukin 1 receptor. *Immunol. Today*, **10**:49–51.

Fink, P. J., Shimonkevitz, R. P. & Bevan, M. J. (1988) Veto cells. *Annu. Rev. Immunol.*, **6**:115–137.

Gastl, G. & Huber, C. (1988) The biology of interferon actions. *Blut*, **56**:193–199.

Geczy, C. L. (1984) The role of lymphokines in delayed-type hypersensitivity reactions. *Springer Semin. Immunopathol.*, **7**:321–346.

Goldstein, G., Scheid, M. P., Boyse, E. A., Schlesinger, D. H. & van Wauwe, J. (1979) A synthetic pentapeptide with biological activity characteristic of the thymic hormone thymopoietin. *Science*, **204**:1309–1310.

Griesser, H., Champagne, E., Tkachuk, D., Takihara, Y., Lalande, M., Baillie, E., Minden, M. & Mak, T. W. (1988) The human T cell receptor α-δ locus: a physical map of the variable, joining and constant region genes. *Eur. J. Immunol.*, **18**:641–644.

Grimm, E. A., Owen-Schaub, L. B., Loudon, W. G. & Yagita, M. (1988) Lymphokine-activated killer cells. Induction and function. *Ann. N. Y. Acad. Sci.*, **532**:380–386.

Groscurth, P., Qiao, B. Y., Podack, E. R. & Hengartner, H. (1987) Cellular localization of perforin 1 in murine cloned cytotoxic T lymphocytes. *J. Immunol.*, **138**:2749–2752.

Hamilos, D. L. (1989) Antigen presenting cells. *Immunol. Res.*, **8**:98–117.

Haynes, B. F., Denning, S. M., Singer, K. H. & Kurtzberg, J. (1989) Ontogeny of T cell precursors: a model for the initial stages of human T cell development. *Immunol. Today*, **10**:87–91.

Henney, C. S. (1989) Interleukin 7: effects on early events in lymphopoiesis. *Immunol. Today*, **10**:170–173.

Hochstenbach, F. & Brenner, M. B. (1989) T-cell receptor δ-chain can substitute for α to form a $\beta\delta$ heterodimer. *Nature*, **340**:562–565.

Hochstenbach, F., Parker, C., McLean, J., Gieselmann, V., Band, H., Bank, I., Chess, L., Spits, H., Strominger, J. L., Seidman, J. G. & Brenner, M. B. (1988) Characterization of a third form of the human T cell receptor γ/δ. *J. Exp. Med.*, **168**:761–776.

Isakov, N., Mally, M. I., Scholz, W. & Altman, A. (1987) T lymphocyte activation: the role of protein kinase C and the bifurcating inositol phospholipid signal transduction pathway. *Immunol. Rev.*, **95**:89–111.

Janeway, C. A. (1989) A primitive immune system. *Nature*, **341**:108.

Janeway, C. A., Jr. (1988) Do suppressor T cells exist? A reply. *Scand. J. Immunol.*, **27**:621–628.

Jenne, D. E. & Tschopp, J. (1988) Granzymes, a family of serine proteases released from granules of cytolytic T lymphocytes upon T cell receptor stimulation. *Immunol. Rev.*, **103**:53–71.

Jung, L. K. L., Fu, S. M., Hara, T., Kapoor, N. & Good, R. A. (1986) Defective expression of T cell-associated glycoprotein in severe combined immunodeficiency. *J. Clin. Invest.*, **77**:940–946.

Kishimoto, T. (1989) The biology of interleukin-6. *Blood*, **74**:1–10.

Koide, J., Rivas, A. & Engleman, E. G. (1989) Natural killer (NK)-like cytotoxic activity of allospecific T cell receptor-γ,δ^+ T cell clones. *J. Immunol.*, **142**:4161–4168.

Kourilsky, P. & Claverie, J.-M. (1989) MHC restriction, alloreactivity, and thymic education: a common link? *Cell*, **56**:327–329.

Kovacs, J. A. & Masur, H. (1988) Opportunistic infections. In V. T. DeVita, Jr., S. Hellman & S. A. Rosenberg, eds., *AIDS: Etiology, Diagnosis, Treatment, and Prevention*, Second Edition, J. B. Lippincott, Philadelphia, pp. 199–226.

Kozbor, D., Trinchieri, G., Monos, D. S., Isobe, M., Russo, G., Haney, J. A., Zmijewski, C. & Croce, C. M. (1989) Human TCR-γ^+/δ^+, $CD8^+$ T lymphocytes recognize tetanus toxoid in an MHC-restricted fashion. *J. Exp. Med.*, **169**:1847–1851.

Larrick, J. W. (1989) Native interleukin 1 inhibitors. *Immunol. Today*, **10**:61–66.

van Leeuwen, B. H., Martinson, M. E., Webb, G. C. & Young, I. G. (1989) Molecular organization of the cytokine gene cluster involving the human IL-3, IL-4, IL-5, and GM-CSF genes, on human chromosome 5. *Blood*, **73**:1142–1148.

Lehner, T. & Brines, R. (1988) Phenotypic and functional characterization of human contrasuppressor cell interactions. *Immunol. Res.*, **7**:33–44.

Low, T. L. & Goldstein, A. L. (1984) Thymosins: structure, function and therapeutic applications. *Thymus*, **6**:27–42.

Lynch, R. G. (1987) Immunoglobulin-specific suppressor T cells. *Adv. Immunol.*, **40**:135–151.

Mahapatro, D. & Mahapatro, R. C. (1984) Cutaneous basophil hypersensitivity. *Am. J. Dermatopathol.*, **6**:483–489.

Marrack, P., McCormack, J. & Kappler, J. (1989) Presentation of antigen, foreign major histocompatibility complex proteins and self by thymus cortical epithethelium. *Nature*, **338**:503–505.

Maury, C. P., Salo, E. & Pelkonen, P. (1989) Elevated tumor necrosis factor-α in patients with Kawasaki disease. *J. Lab. Clin. Med.*, **113**:651–654.

Möller, G. (1988) Do suppressor T cells exist? *Scand. J. Immunol.*, **27**:247–250.

Müllbacher, A. & King, N. J. C. (1989) Target cell lysis by natural killer cells is influenced by β_2-microglobulin expression. *Scand. J. Immunol.*, **30**:21–29.

Müller-Eberhard, H. J. (1988) The molecular basis of target cell killing by human lymphocytes and of killer cell self-protection. *Immunol. Rev.*, **103**:87–98.

Newburger, P. E. & Ezekowitz, R. A. (1988) Cellular and molecular effects of recombinant interferon-γ in chronic granulomatous disease. *Hematol. Oncol. Clin. North Am.*, **2**:267–276.

Ortaldo, J. R. (1986) Comparison of natural killer and natural cytotoxic cells: characteristics, regulation, and mechanism of action. *Pathol. Immunopathol. Res.*, **5**:203–218.

Park, L. S., Friend, D., Sassenfeld, H. M. & Urdal, D. L. (1987) Characterization of the human B cell stimulatory factor 1 receptor. *J. Exp. Med.*, **166**:476–488.

Plate, J. M. D., Lukaszewska, T. L., Bustamante, G. & Hayes, R. L. (1988) Cytokines involved in the generation of cytolytic effector T lymphocytes. *Ann. N. Y. Acad. Sci.*, **532**:149–157.

Quesada, J. R. (1987) Alpha interferons in hairy cell leukemia: a model of biologic therapy for cancer. *Interferon*, **8**:111–134.

Rich, R. R., ElMasry, M. N. & Fox, E. J. (1986) Human suppressor T cells: induction, differentiation, and regulatory functions. *Hum. Immunol.*, **17**:369–387.

Sanders, M. E., Makgoba, M. W. & Shaw, S. (1988) Human naive and memory cells. *Immunol. Today*, **9**:195–199.

Schrieber, L. (1988) Immunomodulators. *Agents Actions (Suppl.)*, **24**:254–264.

Smith, K. A. (1988) The interleukin 2 receptor. *Adv. Immunol.*, **42**:165–179.

Strauss, W. M., Quertermous, T. & Seidman, J. G. (1987) Measuring the human T cell receptor δ-chain locus. *Science*, **237**:1217–1219.

Toyonaga, B. & Mak, T. W. (1987) Genes of the T-cell antigen receptor in normal and malignant T cells. *Annu. Rev. Immunol.*, **5**:585–620.

Umetsu, D. T., Jabara, H. H., DeKruyff, R. H., Abbas, A. K., Abrams, J. S. & Geha, R. S. (1988) Functional heterogeneity among human inducer T cell clones. *J. Immunol.*, **140**:4211–4216.

Vitetta, E. S., Bossie, A., Fernandez-Botran, R., Myers, C. D., Oliver, K. G., Sanders, V. M. & Stevens, T. L. (1987) Interaction and activation of antigen-specific T and B cells. *Immunol. Rev.*, **99**:192–239.

Wilkinson, P. C. & Islam, L. N. (1989) Recombinant IL-4 and IFN-γ active locomotor capacity in human B lymphocytes. *Immunology*, **67**:237–243.

Willems, J., Joniau, M., Cinque, S. & Van Damme, J. (1989) Human granulocyte chemotactic peptide (IL-8) as a specific neutrophil degranulator: comparison with other monokines. *Immunology*, **67**:540–542.

Wodnar-Filipowicz, A., Heusser, C. H. & Moroni, C. (1989) Production of haemopoietic growth factors GM-CSF and interleukin-3 by mast cells in response to IgE receptor-mediated activation. *Nature*, **339**:150–152.

Young, J. D.-E., Liu, C.-C., Persechini, P. M. & Cohn, Z. A. (1988) Perforin-dependent and -independent pathways of cytotoxicity mediated by lymphocytes. *Immunol. Rev.*, **103**:161–202.

Chapter 8

The Major Histocompatibility Complex

The major histocompatibility complex (MHC) is a group of closely linked, highly polymorphic genes. The proteins encoded by these genes are crucial in the generation and execution of immune responses. MHC gene products are necessary for T cell antigen recognition, for collaboration of T cells with accessory cells and B cells in the induction of humoral immune responses, and for cellular immunity against infected cells, or foreign or allogeneic cells (graft rejection). A similar gene complex exists in all vertebrate species yet examined, indicating its conservation over a long period of evolutionary time.

The MHC takes its name from early research indicating the genetic control of the success or failure of organ transplantation. Early clinical observations of skin grafting in humans, and extensive experimental research with animals showed that grafts from one region to another in one individual, or between genetically identical individuals are tolerated, while grafts between genetically disparate individuals are usually rejected. Successfully grafted tissues were "compatible" with their host, while rejected grafts were not. The human MHC is called *HLA* for *human leukocyte antigen*; it resides on chromosome 6 (Figure 8.1). While MHC refers to the gene complex in all species in which it occurs, the term HLA refers specifically to the MHC complex of humans.

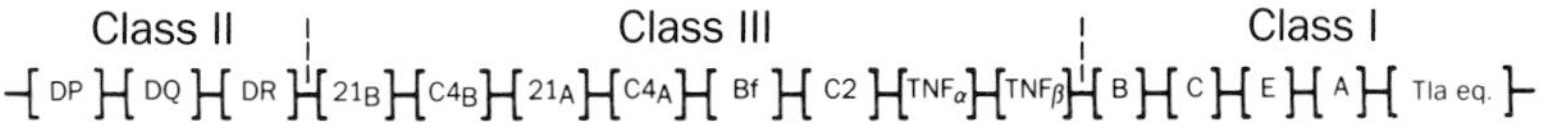

Class II | Class III | Class I

DP | DQ | DR | 21B | C4B | 21A | C4A | Bf | C2 | TNFα | TNFβ | B | C | E | A | Tla eq.

Figure 8.1. *A map of the human MHC (HLA).* The HLA complex resides on chromosome 6. The HLA has not yet been completely mapped, and the nomenclature of certain loci and subregions is in a state of flux. Three subdivisions of class II loci (the D region) are shown here, DP, DQ, and DR. Each contains one or a few α and β genes. Class III loci include complement proteins, 21-steroid hydroxylase, and tumor-necrosis factor (TNF) genes. The class I loci consist of the A, B, and C genes, and the human equivalents of the murine Tla genes, including the E gene, and the F and G genes located 3′ to A (not shown). (Adapted from Carroll et al., 1987.).

Classes and structure of histocompatibility antigens

Class I molecules

The MHC contains two groups of class I antigens distinguished primarily by their tissue distribution. The HLA-A, B and C molecules comprise one group. These antigens are expressed on all cells except erythrocytes and embryonic cells forming the placenta (trophoblast). The surface density varies widely between different cell types. These molecules play an important role in antigen recognition by cytotoxic T lymphocytes, and in graft rejection. Another group of class I molecules has a more restricted distribution, they are found only on some T cells. In mice these are called "Tla" for "thymus leukemia antigen". Three "Tla equivalents" have been found in humans, they are called E, F, and G. The special functions of Tla antigens are not known.

A complete class I molecule consists of two non-covalently linked polypeptide chains (Figure 8.2). The heavy chain with $M_r = 44{,}000$ is encoded by an MHC class I gene. The other chain is called $\beta2$

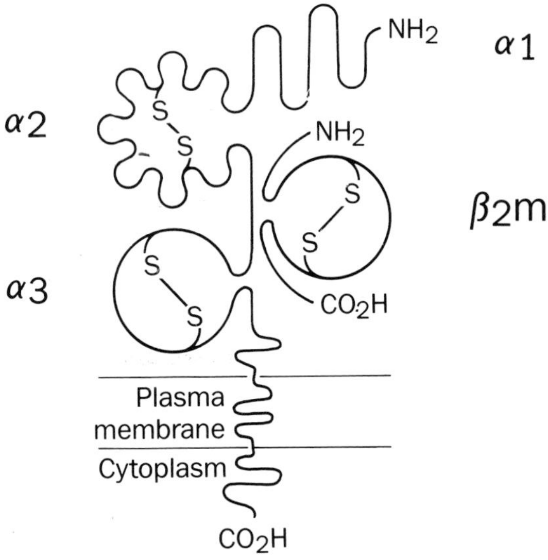

Figure 8.2. *Structure of an MHC class I antigen.* The heavy chain, encoded by an MHC class I gene, has three domains, one of which resembles an immunoglobulin domain. The class I antigen is completed when the heavy chain complexes with a β_2 microglobulin molecule. A peptide binding pocket is formed by two loops of amino acids, one from the α_1 domain, the other from α_2 (Figure 8.6).

microglobulin (M_r = 12,000) and is encoded by a gene not within the MHC (it is on chromosome 2). The A, B, and C loci are *polymorphic*, several different alleles for each of these genes exist in the human species. The gene encoding $\beta 2$ microglobulin is *monomorphic*, only one allele occurs in the species. Thus, all class I antigens contain the same $\beta 2$ microglobulin chain. Class I antigens are not functional in the absence of a $\beta 2$ microglobulin molecule. The heavy chain contains intra-cytoplasmic, transmembrane, and extracellular regions. The extracellular portion has three domains, one of which has structure similar to that of an "immunoglobulin" domain (see Chapter 5).

Class II molecules

Early in the study of these antigens, class II genes were called *immune response* (or *Ir*) *genes* because patterns of immune responsiveness of inbred mice to particular antigens mapped to these loci. As serological tools (antibodies) specific for class II antigens were developed, these MHC antigens became known as *I-region associated* (or *Ia*) *antigens* because they were inherited along with Ir genes.

A class II molecule is a heterodimer of two glycoproteins α (M_r = 34,000) and β (M_r = 29,000), both encoded by MHC genes in the D subregion of the HLA complex (Figure 8.3). This region is divided into three subloci designated DP, DQ, and DR. Each of these loci contains several genes encoding α and β chains. As are class I molecules, class II antigens are integral membrane proteins with intra- and extracellular regions separated by a hydrophobic transmembrane region. The extracellular portion is organized into two domains, one of these has immunoglobulin-like structure.

Human α and β chains appear to show preference for pairing within a subregion. That is, DPα chains are found most often paired with DPβ chains. The extent to which human α and β chains from different subregions pair with one another remains to be determined.

Genes encoding class II antigens are among the most polymorphic known, each having many alleles. These polymorphisms have been defined by allo-antisera (antibodies specific for a particular class II antigen), by mixed lymphocyte culture (see below), and by analysis of DNA restriction fragment length polymorphisms (RFLPs). RFLPs are differences in the pattern of bands detected in a Southern blot when genomic DNA is digested with different restriction endonucleases. Nucleotide sequence analysis has defined portions of the class II genes

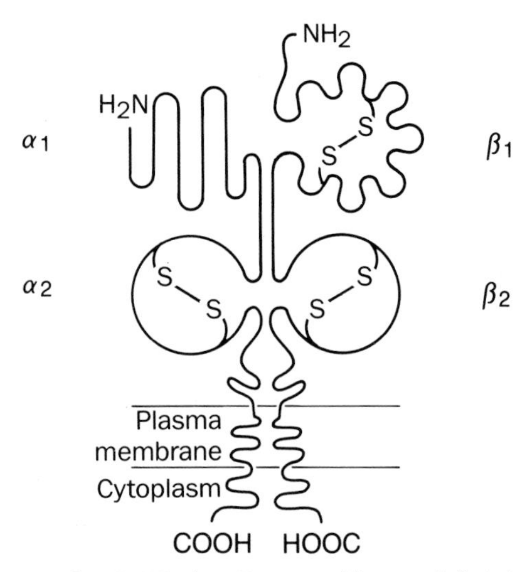

Figure 8.3. *Structure of an MHC class II antigen.* The α and β chains each have two domains, one having structure similar to Ig domains.

in which polymorphic variability is concentrated (hypervariable regions), much as in the variable regions of immunoglobulin or TCR genes.

Class II antigens are expressed mainly by B cells, cells of the monocyte-macrophage lineage (Figure 8.4), and other accessory cells (see Chapter 7), and activated T cells. Cells which do not normally express class II antigens may do so if they are infected by viruses or if they become targets of autoimmune reactions (see Chapter 10). Class II antigens are important for antigen recognition by helper T cells, the collaboration of T cells and B cells (Chapter 7), and in graft rejection.

Class III molecules

MHC class III loci do not encode cell-surface proteins. Rather, they encode some of the components of complement, the enzyme 21-steroid hydroxylase, and tumor necrosis factors (Figure 8.1).

The immunoglobulin gene superfamily

The reader may have noticed some patterns in the structures of immunoglobulins, T cell receptors, and MHC antigens. Various

A

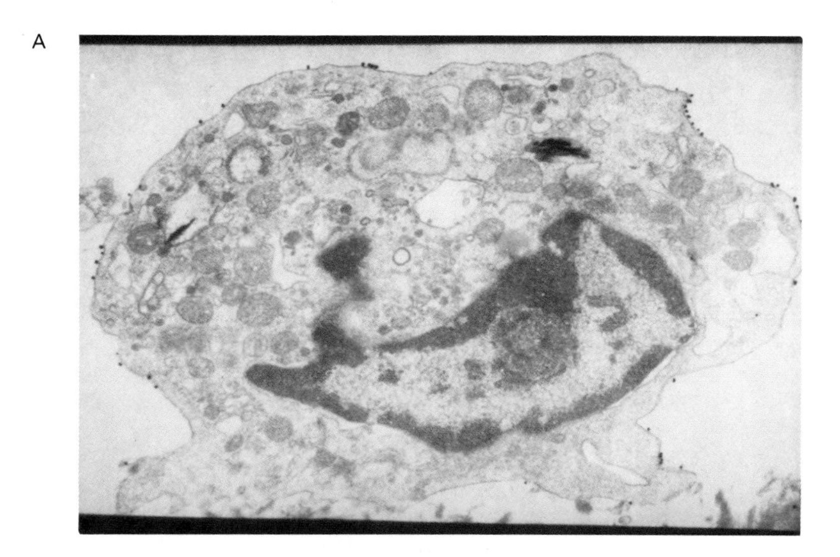

B

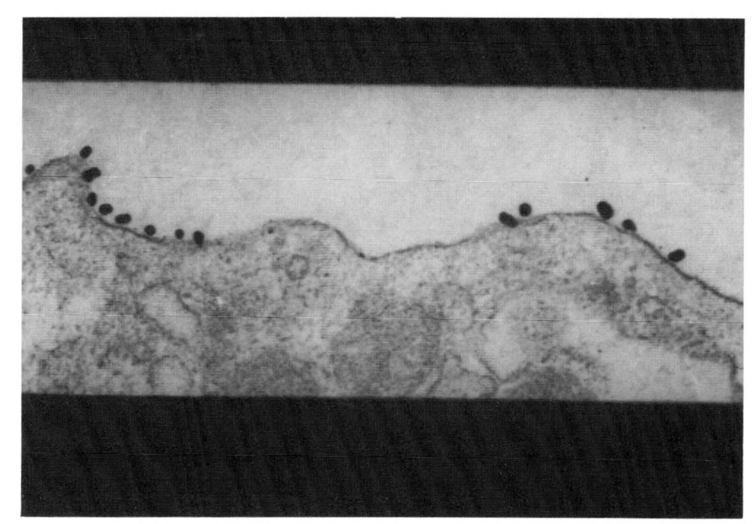

Figure 8.4. *Demonstration of macrophage MHC class II molecules.* A section of a rat skin allograft was incubated with rabbit antibodies against rat MHC class II molecules. The specimen was then incubated with gold-labelled goat antibodies specific for rabbit immunoglobulin.

A. *Perivascular macrophage.* Gold atoms are very electron dense and appear as small black dots in electron micrographs. These are clearly visible distributed over the surface of this macrophage.

B. *Detail of A.* A portion of panel A has been enlarged to show more clearly the staining of the macrophage surface.

Table 8.I. THE IMMUNOGLOBULIN GENE SUPERFAMILY

Immunoglobulins, heavy and light chains
T cell receptor, α, β, γ, δ
CD3 complex γ, δ, ε
MHC class I heavy chain, class II α and β, and β2 microglobulin
CD1a heavy chain
CD2
CD4
CD8
CD58
Receptors for: IgG Fc, platelet-derived growth factor, colony stimulating factor-1
Carcinoembryonic antigen

(Adapted from Williams and Barclay, 1988.)

domains of the latter groups of proteins have been likened to those of immunoglobulin molecules. Many more such proteins have been found, and the genes encoding them are found widely dispersed in the genome. These genes, because of their structural relatedness, may be considered members of an *immunoglobulin gene superfamily*, and are thought to have derived from one or a few "primordial" immunoglobulin-like genes. Table 8.I is a partial list of the proteins whose genes have been induced in this family. Because all of these molecules are integral membrane proteins, they are presumed to have some role in cellular interactions, or as receptors.

MHC specificities and haplotypes

We have mentioned that MHC class I and II loci are very polymorphic. The total number of alleles for all of these loci is not known, but is probably on the order of hundreds. The proteins encoded by different alleles of a particular MHC gene differ slightly from one another in structure. Prior to the advent of recombinant DNA technology, MHC genetics was explored with antibodies specific for their products, the MHC antigens. Human anti-HLA antibodies may be obtained from multiparous women producing antibodies against paternal MHC antigens encountered during pregnancy, or from individuals who have received multiple blood transfusions. For this reason, one will often find MHC antigens described as *MHC specificities*, a terminology resulting from production of an antibody with that particular "specificity."

Another powerful method for studying the MHC is *mixed lymphocyte culture* (*MLC*), also called the *mixed lymphocyte reaction* (*MLR*). Lymphocytes from genetically different individuals proliferate when mixed and cultured together. In performing an MLR experiment, one set of lymphocytes is designated *stimulator cells*, the other, *responder cells*. The stimulator cells are rendered incapable of dividing (but not killed) either by irradiation or by treatment with a chemical such as mitomycin C. The stimulators and responders are then mixed, and a radioactively labelled nucleotide (e.g., ^{3}H-thymidine) is added to the culture medium. Proliferation of responder cells is easily determined by measuring the amount of radioactivity that has been incorporated. Differences in both MHC class I and class II antigens may determine proliferation in MLR. In general, differences in class I loci result in weak MLR, while differences in class II loci give strong reactions. Monocytoid cells are much more potent stimulators than are lymphocytes. The cells which proliferate in MLR are *alloreactive* T cells. These are T cells which are activated by foreign MHC molecules on APC or lymphocyte surfaces.

The nomenclature of HLA alleles or specificities follows simple rules. The designation contains up to three components: the locus (A, B, D, DR, etc.), the letter "w" standing for the word "workshop" to indicate a specificity awaiting confirmation, and a number. Thus, A1, Aw43, B12, Bw53, Cw1, DR3, etc. Note that specificities of the C locus always contain the letter "w" to distinguish them from complement components.

It is often useful to consider a group of closely linked loci as a unit in comparisons between individuals or populations defined by some characteristic. This is the basis of the concept of *haplotype*. The haplotype of a group of loci is a designation describing the alleles present at each locus. An HLA haplotype is designated simply by writing consecutively the antigens at each locus. An example of a haplotype might be: A1, B8, Cw7, DR3 (not all loci need be specified).

A haplotype refers only to a gene complex on one chromosome. Since somatic cells are diploid, two haplotypes are required to describe an individual's total genetic information at a particular group of loci. It is important to keep in mind that MHC antigens are co-dominantly expressed. Therefore, the MHC antigens corresponding to both haplotypes will be equally abundant. With such a large pool of available HLA alleles, it is very unlikely that a member of a relatively outbred species such as humans will be homozygous at all HLA loci. It is also highly unlikely that two individuals chosen at random will have even one identical haplotype, much less two. From one pair of parents, four haplotype combinations are possible. Thus, among five full siblings, at least two must have identical HLA types, unless recombination has occurred within the HLA complex during meiosis.

Although humans are relatively outbred, they are not completely so. It is only quite recently in history that the technology of travel has permitted rapid intermingling of previously isolated populations. This aspect of human history is evident in the frequencies with which particular HLA alleles occur in various ethnic groups. Some alleles such as A2 are found with similar frequency in all groups. Others show marked restrictions. For example, A1 and A3 are present in 15.8% and 12.6% of European caucasians, respectively. These specificities are virtually absent from Japanese.

In a randomly mating population at equilibrium, the probability of two particular alleles occurring together is the product of their frequencies in the gene pool. However, the frequency with which particular MHC specificities (genes) occur together may be very different from what would be expected from their individual frequencies in a population. When the observed frequency of a gene combination

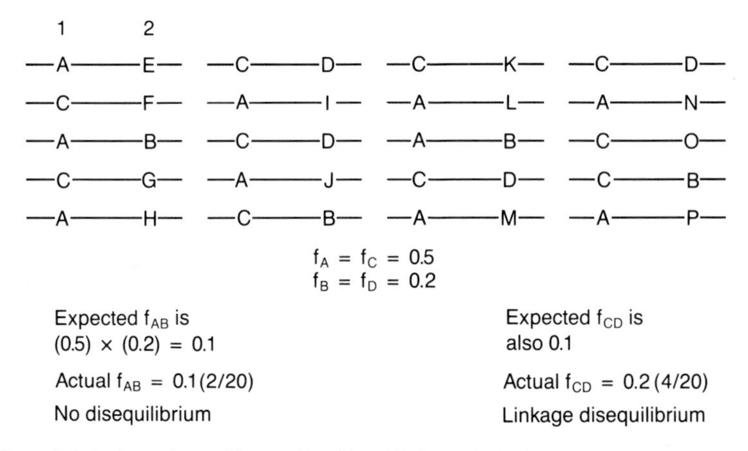

Figure 8.5. *Linkage disequilibrium.* Consider this hypothetical MHC with two gene loci, 1 and 2. Only two alleles occur in locus 1. A and C. Many more alleles are found in locus 2, these are B and D-P. The frequencies of Alleles A and C are equal, each being 0.5 (10/20 haplotypes in this example). The frequencies of alleles B and D are also equal, each being 0.2 (4/20 haplotypes). The expected frequency of any two alleles occurring together is equal to the product of their individual respective frequencies. For the association of A with B, and that of C with D, this frequency is equal to 0.1 (0.5 × 0.2) in both cases. The frequency of A with B in this population is equal to that expected, however, C is found associated with D twice as often as expected. This constitutes linkage disequilibrium.

is markedly different (higher or lower) than the predicted frequency, the two alleles are said to be in *linkage disequilibrium* (Figure 8.5). For example, in caucasians, A1 and B8, or A3 and B7, or B8 and DRw3 occur together more frequently than expected based on the individual frequencies of A1, B8, A3, B7, and DRw3 in caucasians.

Linkage disequilibrium may reflect some selective force acting either in favor of or against a particular combination of alleles. Selective forces influencing the frequencies of MHC haplotypes and linkage of particular loci may act only on MHC genes, may operate only on other genes linked to MHC, or (probably) both. Perhaps the combinations of alleles in certain haplotypes characterize more "effective" immune systems. Alternatively, linkage disequilibrium may reflect the fact that the human species is not completely outbred.

Functions of MHC antigens

Genetic restriction

As we have described previously, T cells only recognize antigens when they are associated with MHC molecules. In addition, several interactions between T, B, and other cell types, important in generating immune responses, are dependent on MHC gene products. This function of MHC molecules in antigen recognition and cellular cooperation in immune responses is called *genetic restriction* since these interactions are restricted to cells expressing compatible MHC antigens and receptors.

Different T cell subsets recognize antigens in association with different MHC molecules. Helper/inducer, delayed hypersensitivity, and contact sensitivity T cells (all $CD4^+$) recognize antigen associated with class II gene products. Interactions of T cells with B cells are also restricted by class II antigens. Cytotoxic T cells ($CD8^+$) are activated by antigen complexed with class I molecules. *Allospecific* or *alloreactive* T cells lyse cells bearing foreign MHC class I antigens. A subset of alloreactive cytotoxic cells express CD4 and lyse cells bearing foreign MHC class II antigens. Both of these cell types are evident in MLR, graft rejection, and graft-versus-host phenomena (see below). Recall from Chapter 7 that genetic restriction is programmed by the thymic microenvironment in which T cells mature, not the MHC antigens which they themselves encode.

Antigen association with MHC molecules as an intermediate step in T cell antigen recognition and activation is called *antigen presentation*

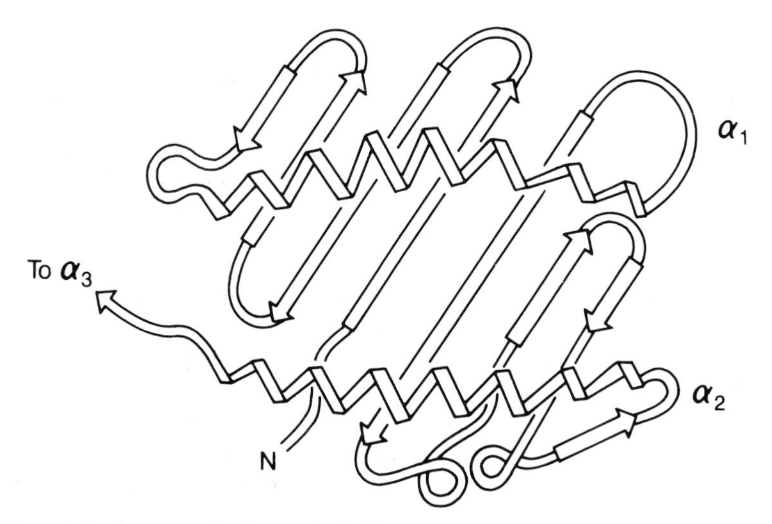

Figure 8.6. *The antigen-binding pocket of an HLA class I molecule.* This is a schematic representation of the structure of HLA-A2 as determined by X-ray crystallography. The view is "downward" looking at the top of the molecule as it would sit in the plasma membrane. Long parallel alpha helices from the α_1 and α_2 domains form a binding pocket that quite resembles a mouth. This pocket is the receptor structure which accommodates antigenic peptide fragments. (Adapted from Bjorkman et al., 1987.)

(Chapter 7). The class I or II molecules of a given haplotype bind a large number of structurally diverse peptide antigen fragments. Thus, they are relatively nonspecific peptide receptors. Crystallographic studies of the class I HLA-A2 antigen showed that the domains fold together to create a binding pocket (Figure 8.6). Almost all of the residues in variable (polymorphism-determining) regions surround the binding pocket. Although not yet demonstrated directly, class II MHC molecules are also thought to have a similar binding pocket. The molecular basis of the genetic restriction of antigen recognition, then, is the insertion of peptide antigen fragments into the binding pocket of MHC class I and class II molecules.

The MHC and immune responsiveness

Since MHC molecules are so important in immune recognition of the majority of protein and glycoprotein antigens, it is not surprising that genes mapping to MHC loci are associated with differences in the magnitude and quality of both humoral and cellular immune responses to a number of antigens. These "immune response" (Ir) genes map to the same locus determining strong reactions in mixed lymphocyte culture, the MHC class II locus. In examining the response to a particular antigen, it is possible to distinguish one or a few specificities associated with a positive response. For example, response to *S. japonicum* is associated with DR2, and response to *P. falciparum* circumsporozoite protein with DRw13. The lack of a response may also be associated with specific class II genes. In the examples cited above, non-response to the schistosome is associated with DQw1, and non-response to the malarial antigen with DQw3. This "immune suppressive" or "Is" gene phenomenon may be due to antigen-specific suppressive mechanisms associated with expression of particular class II molecules. This is supported by experiments showing that when DR2 and DQw1 (the determinants of response and non-response to schistosomal antigens, respectively) are present in the same individual, that the non-responder phenotype is expressed. The "suppressor" gene is able to eliminate the positive effect of the responder gene.

Ir genes not only determine response or non-response to a particular antigen, but they may also influence the quality of a response. An individual possessing a particular MHC haplotype may produce antibodies of higher affinity for an antigen, or of different isotypes than another individual with a different haplotype.

An early experimental finding which perplexed immunologists was that the Ir phenotype with respect to a humoral response appeared to be determined by T cells, but not B cells. If an irradiated mouse is reconstituted with B cells from a non-responder, and T cells from a responder, the mouse will be a responder. However, if the mouse is reconstituted with T cells from a non-responder, and B cells from a responder, the mouse will be a non-responder. The results of this kind of experiment are shown in Table 8.II. This phenomenon results from

Table 8.II. ASSOCIATION OF THE RESPONDER PHENOTYPE WITH T CELLS

Irradiated mice reconstituted with		Antibody
B cells from:	T cells from:	response
non-responder	—	−
responder	—	−
—	non-responder	−
—	responder	−
non-responder	non-responder	−
responder	non-responder	−
non-responder	responder	+
responder	responder	+

the genetic restriction of the interaction between helper T cells and antigen presenting cells. This interaction is the key step in an immune response that is influenced by Ir gene products (class II MHC antigens).

It may be supposed that the association of the responder phenotype with particular class II specificities means that these class II molecules are able to bind the foreign antigen or its fragments. Conversely, non-responders do not have class II molecules capable of associating with the antigen. This is known as the *determinant selection hypothesis.* An alternative explanation is that there are no T cells in non-responders capable of being activated by that particular combination of antigen and class II molecules. This might occur if the individual was actually incapable of creating a T cell receptor with that antigen specificity, or if such T cells were destroyed in the thymus (*hole in the repertoire hypothesis*). Alternatively, such T cells may exist, but they are somehow rendered incapable of activation, perhaps by suppressor T cells (*blind spot hypothesis*). There is experimental evidence to support each of these explanations for Ir gene phenomena. Perhaps all may operate in different circumstances.

Transplantation

The vocabulary of transplantation is quite redundant, but follows simple patterns. The prefixes *auto-*, *homo-*, *iso-*, and *syn-* mean "self" or "same" or "with." Thus, the adjectives *autogenous*, *autologous*, *homologous*, *isogeneic*, *isologous* and *syngeneic* all describe grafts from one site to another in one individual, or between two genetically identical individuals. The related nouns are *autograft*, *homograft*, and *isograft*. The prefix *allo-* means "other," thus, *allogeneic* and *allograft* describe transplantation between two genetically different individuals of the same species. The prefix *hetero-* also means "other," while *xeno-* means "foreign." Consequently, *heterologous*, *heterospecific* and *xenogeneic*, together with *heterograft* and *xenograft* describe the transfer of tissue across species lines. More synonyms for these words exist, however, we hope the general pattern of nomenclature is sufficiently evident in the above selection.

Syngeneic grafts are normally not destroyed by the immune system. Since the MHC antigens expressed in the graft are those existing throughout the body, they are simply components of "self" and do not elicit an immune response. In addition to autografts and syngeneic grafts, transplants from two different inbred (homozygous) parents to F_1 progeny are not rejected. This is because the MHC antigens corresponding to each of the parents' haplotypes is expressed in the F_1 recipient. On the other hand, grafts from the F_1 to the parents are rejected since one parent does not express the MHC antigens which the F_1 inherited from the other.

Allografts and heterografts are usually rejected by the immune system of the recipient. The MHC antigens in these grafts are different from those regarded as self, and elicit vigorous cell-mediated and humoral immune responses leading to graft destruction. Initiation of graft rejection is T cell-dependent.

Graft rejection may be either acute or chronic. The rapidity with which a graft is destroyed depends on the degree of dissimilarity between the MHC antigens of the host and donor, whether or not the host received grafts from the same or genetically (MHC) similar individuals in the past, and whether or not the immune system of the recipient has been pharmacologically or pathologically suppressed.

Graft rejection is initiated by lymphocytes reacting to MHC antigens expressed in the graft. Helper T cells secrete soluble factors (cytokines, see Chapter 7) activating cytotoxic T cells and B cells and recruiting more lymphocytes, as well as many mononuclear cells and neutrophils. The graft is destroyed both by specific and non-specific cellular cytotoxicity, as well as antibody-dependent cellular cytotoxicity (see Chapter 7) and complement-mediated lysis (Chapter 4).

The principal mechanism of graft rejection is inflammation. The earliest histological changes in the graft undergoing rejection are vascular dilation and increased permeability leading to diffuse leakage of blood. In a vigorous, acute rejection (low histocompatibility), the graft is progressively infiltrated by lymphocytes and macrophages and eventually undergoes necrosis. Rejection of large grafts may also be accompanied by systemic symptoms such as malaise, anorexia, fever, and myalgia. Chronic rejection (high histocompatibility) is similar, but less severe. Mild inflammation and less cellular infiltration is observed. Bleeding and necrosis are less pronounced.

Rejection of the first graft from a particular donor is called *first set rejection*. Rejection of the second graft from the same donor is called *second set rejection*. In general, second set rejection is much more rapid than first set rejection. This phenomenon reflects the priming of the immune system to the MHC antigens encountered in the first graft. Second set rejection is an example of a secondary immune response (see Chapter 7).

MHC antigens are not the only determinants of histocompatibility, they are simply the *major* determinants. An unknown number of (possibly several hundred) poorly characterized *minor histocompatibility antigens* exist. Identity at MHC loci with differences at minor loci may result in very slow (but nevertheless complete) graft rejection over a period of weeks or months. ABO blood group antigens are also found on other cells and constitute important transplantation antigens.

How does one decide who will be the most compatible recipient of a particular organ? Blood samples from potential donors and recipients are subjected to both serological and cytological tests. Screening with standardized monoclonal antibodies and human antisera may determine many class I and class II specificities. MLR and several variations on this technique may reveal additional specificities.

The most commonly used means for prolonging graft survival is suppression of the recipient's immune system with drugs (e.g., cyclosporine). Clearly, this is not an ideal solution since it leaves the patient less able to cope with infection. Much research is currently directed toward developing more specific means of preventing or suppressing graft rejection.

The graft versus host reaction

Bone marrow transplantation may be curative of some leukemias and blood dyscrasias. The patient's hemopoietic ability is completely destroyed by irradiation and/or cytotoxic chemotherapy prior to reconstitution with bone marrow from a donor. In this situation, immunocompetent cells in the bone marrow graft may generate a response against antigens in the host. This is called a *graft-versus-host (GVH) reaction.* Mature T cells in the graft are activated by major and minor histocompatibility determinants, proliferate and secrete cytokines generating a tissue-destructive reaction. In the latter respect, both graft rejection and GVH reactions are similar to the delayed hypersensitivity reaction (Chapter 7). Bone marrow grafts are most successful between HLA-identical siblings; they are rarely successful between unrelated individuals with even minor histoincompatibility.

As may graft rejection, a GVH reaction may be acute or chronic. The principal organs involved are the skin, liver, and intestines. These organs undergo mild or severe inflammation and necrosis. Skin may desquamate, liver function fail, and malabsorption occur. Additional symptoms include weight loss, lymphadenopathy, hepatosplenomegaly, and increased susceptibility to infection. Death invariably results without treatment. Current therapies use immunosuppressive drugs to quell the reaction. Even without such immunosuppression, the cause of death is most often overwhelming infection. One promising new possibility for preventing GVH disease is depletion of mature T cells in the grated marrow. The marrow then contains no cells capable of proliferating in response to the host MHC antigens, and the T cells which subsequently develop in the host thymus from donor precursors will be tolerant to host antigens.

MHC AND DISEASE

Diseases associated with specific haplotypes

A number of diseases having an inherited component of susceptibility or etiology are linked to MHC haplotypes. Such an association is defined, simply enough, by comparing the frequency of an MHC allele or haplotype in a healthy population with the frequency of that allele or haplotype in people with a particular disease. One calculates the *relative risk* of developing a disease in a group of people expressing MHC antigen or haplotype X compared to people not expressing X according to the formula:

$$\text{Relative risk} = \frac{p^{+} \times c^{-}}{p^{-} \times c^{+}}$$

Where:

p^{+} is the frequency of patients with disease and haplotype X
p^{-} is the frequency of patients without haplotype X
c^{+} is the frequency of controls (without disease) with X
c^{-} is the frequency of controls without X

Calculating an accurate relative risk may be quite difficult. Many diseases have multifactorial genetic components influencing susceptibility. These components may each have variable linkage with MHC. In addition, many environmental and life style factors also influence disease. Table 8.III lists several diseases which have been linked to MHC. It should be emphasized, however, that this does not exclude a

Table 8.III. HLA-ASSOCIATED DISEASES IN CAUCASIANS

Disease	Associated specificity	Relative risk
Rheumatoid arthritis	DR4	3.8
Ankylosing spondylitis	B27	69.1
Reiter's disease	B27	37.1
Juvenile diabetes mellitus	DR4	3.6
Grave's disease	B35	4.4
Celiac disease	DR3	11.6
Narcolepsy	DR2	129.8
Psoriasis vulgaris	CW6	7.5
Pemphigus vulgaris	DR4	14.6
Dermatitis herpetiformis	DR3	17.3
Behcet's disease	B5	3.8
Idiopathic hemochromatosis	A3	6.7
Sjögren's syndrome	DW3	5.7
Systemic lupus erythematosus	B8	2.7
Goodpasture's syndrome	DR2	13.8
Multiple sclerosis	DR2	2.7
Myasthenia gravis	B8	3.3

Adapted from Tiwari and Terasaki, 1985, p. 33.

role for other genes mapping outside the MHC locus, or environmental factors. For many MHC-linked diseases there is only a 30–70% concordance rate between monozygotic twins.

The most logical first step towards understanding correlations of MHC haplotypes and susceptibility to disease is to examine the structural characteristics of the particular MHC genes/antigens involved.

Rheumatoid arthritis

Rheumatoid arthritis has been associated with DR4 and DR1 in several ethnic groups. White, oriental and black North American RA patients have a high frequency of DR4. Asian Indians with RA have a high frequency of DR1. MLR analysis has subdivided the DR4 specificity into Dw4, Dw10, Dw13, Dw14, and Dw15. An examination of these DR specificities in various ethnic groups has shown significant differences in their association with RA (Table 8.IV).

Sequence analysis of these class II antigens showed identity in the first and second hypervariable regions, and several differences in the third hypervariable region of DRβ1. In particular, the Dw4, Dw14, and

Table 8.IV. ASSOCIATION OF DR SPECIFICITIES WITH RHEUMATOID ARTHRITIS IN VARIOUS ETHNIC GROUPS

Specificity	American caucasians	Japanese
Dw4	+	−
Dw14	+	−
Dw15	−	+

(Data from Gregersen et al., 1988.)

Dw15 genes encode beta chains very similar to that found in DR1. The dual association of RA with DR4 and DR1, then, is presumably due to the similarity of the third hypervariable region of the DRβ1 chains of these class II molecules.

Insulin-dependent (type I) diabetes mellitus (IDDM)

Type I diabetes mellitus has a prevalence of 0.2–0.3% in the USA. The disease is caused by destruction of the β cells of the pancreatic islets of Langerhans with ensuing loss of insulin secretion. The presence of antibodies reactive with β cells and insulin and infiltration of pancreatic islets by T cells in a large fraction of patients has caused much speculation that this is a primary autoimmune disease.

A striking correlation has been found with IDDM and particular amino acids at position 57 in the DQβ chain. Alleles encoding Ala, Val, or Ser at this position are found in high frequency in patients with IDDM, while Asp at this position occurs very rarely. In a murine model for IDDM, the non-obese diabetic (NOD), mouse, the I-Aβ chain (the homologue of DQβ) has Ser at position 57, in contrast to the non-obese normal (NON) strain which has Asp at this position.

Celiac disease (CD, gluten enteropathy)

This disease is a malabsorptive state resulting from lymphoid infiltration of the small intestines with villous atrophy and destruction. The disease presents clinically as severe diarrhea and malabsorption. The diagnosis is made if symptoms remit after removing all gluten (proteins found in wheat and other grains) from the diet. A vigorous immune response to gluten may be the cause of this disease; anti-gluten antibodies are found in most patients.

The DR3-DQw2 haplotype is very frequent in patients with celiac disease. Let us designate the DQα and DQβ genes of this haplotype as DQα_{CD} and DQβ_{CD}. These genes are also found in other haplotypes. DQα_{CD} occurs in DR5-DQw7, while DQβ_{CD} is found in DR7-DQw2. Thus, the DQα/β_{CD} heterodimer may be formed in individuals having the DR3-DQw2 haplotype on one chromosome, or in individuals who are DR5-DQw7/DR7-DQw2 heterozygotes. The observation that 93/94 CD patients have one of these genotypes strongly suggests that this particular DQα/β heterodimer is important in the etiology of this disease.

The mechanisms translating these structural patterns into disease susceptibility remain to be discovered. One possibility is that self tolerance may be more easily disrupted by particular combinations of self or foreign antigens and class II gene products. In particular, viral infections, and cross-reactivity (structural similarity) between foreign and self-antigens are frequently invoked as potential mechanisms underlying autoimmunity (see Chapter 10).

Congenital defects of MHC antigens

The bare lymphocyte syndrome

This disease entity was initially described in an infant lacking MHC class I antigens on lymphocytes and platelets. Among approximately 25 patients studied to date, great heterogeneity in the degree of deficiency of class I or class II antigens has been observed. Since mRNA encoding these MHC antigens is present in lymphocytes of patients with this disease, the underlying defect must be in a separate gene locus influencing expression of MHC antigens on the cell surface.

The number of T cells is less than normal. Normal numbers of B cells are present, but they appear to be arrested at an immature stage of development and are incapable of becoming plasma cells. Hence, the level of serum antibodies is quite low. Depending on the degree of lack of MHC antigens, the symptomatology of the bare lymphocyte syndrome may be minimal, or may mimic severe combined immunodeficiency (SCID, see Chapter 2).

Recently, an attempt was made to reconstitute immune function in a fetus diagnosed with bare lymphocyte syndrome following sampling of cord blood at 21 weeks gestation. Cadaveric fetal liver cells and thymic epithelial cells were injected in the umbilical cord at 30 weeks gestation. There were no problems associated with rejection or GVH phenomena due to the immaturity of grafted cells and the recipient. Although complete immune function was not reconstituted even at seven months after birth, the child's immune function was still superior to that which had been achieved previously with postnatal transplants of fetal tissue. *In utero* transplantation may prove to be a promising method of correcting hemopoietic abnormalities.

Having described in the preceding chapters the mechanisms by which specific immune responses are generated and controlled, one may question what prevents our immune systems from responding to our own antigens. The ability to respond quickly and vigorously to exogenous antigens while remaining quiescent in a milieu abounding with immunogenic molecules is the fundamental dilemma faced by our immune systems. In the next chapter we examine the mechanisms maintaining the sometimes uneasy peace between the immune system and other tissues of the body.

SOURCES AND SUGGESTED ADDITIONAL READING

Baur, M. P. & Danilovs, J. A. (1980) Population analysis of HLA-A, B, C, DR and other genetic markers. In P. I. Terasaka ed., *Histocompatibility Testing 1980*, UCLA Tissue Typing Laboratory, Los Angeles, pp. 955–963.

Bjorkman, P. J., Saper, M. A., Semraoui, B., Bennet, W. S., Strominger, J. L. & Wiley, D. C. (1987) Structure of the human class I histocompatibility antigen, HLA-A2. *Nature*, **329**:506–512.

Bodmer, W. F. & Bodmer, J. G. (1978) Evolution and function of the MHC system. *Br. Med. Bull.*, **34**:309–316.

Carroll, M C., Katzman, P., Alicot, E. M., Koller, B. H., Geraghty, D. E., Orr, H. T., Strominger, J. L. & Spies, T. (1987) Linkage map of the human major histocompatibility complex including the tumor necrosis factor genes. *Proc. Natl. Acad. Sci. USA*, **84**:8535–8539.

Clement, L. T., Plaeger-Marshall, S., Haas, A., Saxon, A. & Martin, A. M. (1988) Bare lymphocyte syndrome. Consequences of absent class II major histocompatibility antigen expression for B lymphocyte differentiation and function. *J. Clin. Invest.*, **81**:669–675.

DeGroot, A. S., Johnson, A. H., Maloy, W. L., Quakyi, I. A., Riley, E. M., Menon, A., Banks, S. M., Berzovsky, J. A. & Good, M. F. (1989) Human T cell recognition of polymorphic epitopes from malaria circumsporozoite protein. *J. Immunol.*, **142**:4000–4005.

Dupont, B. (1989) Nomenclature for factors of the HLA system, 1987. *Hum. Immunol.*, **26**:3–14.

Gregersen, P. K., Silver, J. & Winchester, R. J. (1988) Genetic susceptibility to rheumatoid arthritis and human leukocyte antigen class II polymorphism. The role of shared conformational determinants. *Am. J. Med.*, **85**(6A): 17–19.

Hirayama, K., Matsushita, S., Kikuchi, I., Iuchi, M., Ohta, N. & Sasazuki, T. (1987) HLA-DQ is epistatic to HLA-DR in controlling the immune response to schistosomal antigen in humans. *Nature*, **327**: 426–430.

Joysey, V. C. & Wolf, E. (1978) HLA-A, -B, and -C antigens, their serology and cross-reaction. *Br. Med. Bull.*, **34**: 217–222.

Klein, J. (1986) *Natural History of the Major Histocompatibility Complex*, John Wiley & Sons, Inc., New York.

Medawar, P. B. (1957) The immunology of transplantation. *Harvey Lect.*, **52**: 144–176.

Sollid, L. M., Markussen, G., Ek, J., Gjerde, H., Vartdal, F. & Thorsby, E. (1989) Evidence for a primary association of celiac disease to a particular HLA-DQ α/β heterodimer. *J. Exp. Med.*, **169**: 345–350.

Spies, T., Blanck, G., Bresnahan, M., Sands, J. & Strominger, J. L. (1989) A new cluster of genes within the human major histocompatibility complex. *Science*, **243**: 214–217.

Thomson, G., Robinson, W. P., Kuhner, M. K., Joe, S., MacDonald, M. J., Gottschall, J. L., Barbosa, J., Rich, S. S., Bertrams, J., Baur, M. P., Partanen, J., Tait, B. D., Schober, E., Mayr, W. R., Ludvigsson, J., Lindblom, B., Farid, N. D., Thompson, C. & Deschamps, I. (1988) Genetic heterogeneity, modes of inheritance, and risk estimates for a joint study of caucasians with insulin-dependent diabetes mellitus. *Am. J. Hum. Genet.*, **43**: 799–816.

Tiwari, J. L. & Terasaki, P. I. (1985) *HLA and Disease Associations*, Springer-Verlag, New York.

Touraine, J. L., Raudrant, D., Royo, C., Rebaud, A., Roncarolo, M. G., Souillet, G., Philippe, N., Touraine, F. & Bétuel, H. (1989) In-utero transplantation of stem cells in bare lymphocyte syndrome. *Lancet*, **i**: 1382.

Williams, A. F. & Barclay, A. N. (1988) The immunoglobulin superfamily—domains for cell surface recognition. *Annu. Rev. Immunol.*, **6**: 381–405.

Chapter 9

Tolerance

We do not normally generate immune responses against the multitude of antigens within our bodies, yet we may respond vigorously to antigens in our environment and from other individuals which may differ only subtly from our own. Our immune systems "tolerate" the many immunogens comprising us.

Tolerance is defined as a condition of *antigen-specific immunologic unresponsiveness*. Tolerance must be distinguished from the non-specific unresponsiveness of immunodeficiency, the antigen-specific unresponsiveness determined by Ir genes (see Chapter 8), or the transient antigen unresponsiveness induced by continual administration of large amounts of antigen (*pseudotolerance* or *immunologic paralysis*).

Tolerance may be induced in a variety of ways. Proteins in the foods we eat may be found in small amounts in our blood. In this way, animals may become sensitized to food proteins so that they develop an allergic reaction when they are administered intravenously. Interestingly, animals do not have such a reaction to food proteins which constitute a part of their diet from birth. In some way, lifelong contact with these proteins prevents sensitization leading to allergy. To put it another way, animals are tolerant to the antigens in their customary foods.

Similar phenomena have been observed in experiments with a synthetic antigen, 2:4 dinitrochlorobenzene (DNCB). Allergy to this chemical may be induced in guinea pigs by intradermal injection. If animals are fed DNCB, they do not subsequently become allergically sensitized by injection. Furthermore, tolerance can only be induced by feeding if it is the animal's first contact with the antigen. Animals already sensitizied remain allergic after they are fed DNCB.

Fraternal twin calves occasionally have circulating erythrocytes derived from the other twin. Vascular anastomoses in the placental circulations of the twins permit exchange of mature blood cells and stem cells. Blood cells derived from the twin's stem cells persist in the circulation. Thus, such calves are not only tolerant to the twin's blood, but also to skin grafts. Apparently, exposure to antigens in embryonic development and their persistence in life results in immunologic tolerance.

This conclusion is strengthened by the following observation: newborn strain A mice injected with spleen cells from strain B, will accept skin grafts from strain B as adults. This demonstrates tolerance induced by neonatal injection of cells. The specificity of this tolerance is evidenced by the ability of the strain A mice to reject skin grafts from strain C.

Factors affecting tolerogenicity

Age

Tolerance is most easily induced during embryonic development and during the perinatal period, when the immune system is immature. Tolerance induced at these times usually persists much longer than when induced in adults. B cells from neonates, or those that have just begun to express surface IgM, or recently stimulated memory cells are the most susceptible to tolerization. Although mature resting B cells are relatively resistant to tolerization, they are not absolutely so. They require much higher concentrations of antigen in order to be tolerized. Whether or not there are differences in the mechanism of tolerization of resting versus cycling B cells is not known.

Route of antigen administration

The manner in which the immune system encounters an antigen may affect tolerance induction. As described above, oral administration is frequently tolerogenic. Injection into the portal circulation also tends to be tolerogenic. The similarity of both of these routes of antigen administration is apparent. Intrathymic antigen injection is also tolerogenic. This must be a result of disturbing T cell maturation in the thymus, which is normally free of exogenous antigens.

Physical/chemical nature of the antigen

Animal studies of tolerance to foreign serum proteins showed that monomeric molecules induce tolerance at low doses, while aggregated proteins do not. This may be related to the ease of phagocytosis and presentation of aggregated molecules, and persistence of monomeric forms in the circulation. Some discrepancies may be noted when tolerization is examined *in vitro* and *in vivo*. Parenterally administered monovalent proteins may be good tolerogens *in vivo*, but not *in vitro*.

Molecular size is also important in tolerization. A decrease in molecular size that does not affect antigenicity is accompanied by a decrease in immunogenicity, and an increase in tolerogenicity. For example, polymerized flagellin (a protein component of bacterial flagellae) of $M_r = 10^7$ is a better immunogen than monomeric flagellin ($M_r = 40{,}000$). Fragment A of flagellin ($M_r = 18{,}000$) is more tolerogenic than immunogenic. In general, T-independent antigens are both less immunogenic and less tolerogenic as they become smaller. In some cases, however, there is a discrepancy between these two properties. For example, fructans (fructose polymers) below $M_r = 3$–$6{,}000$ are tolerogenic, but are not immunogenic.

Substances metabolized slowly, or not at all, may induce a type of tolerance which has been called *immunologic paralysis* or *pseudotolerance*. This condition is transient and may occur after administration of large amounts of antigen which is not easily eliminated from the system. For example, type III pneumococcal polysaccharide, or D-amino acid polymers, cannot be hydrolyzed because most higher organisms lack appropriate enzymes. Antigen is internalized by phagocytic cells, but cannot be degraded and is eventually exocytosed. Thus, the antigen may persist in high concentration in the circulation. While antigen persists, the animal is unable to generate an immune response. This is believed to result from saturating B cell surface Ig with antigen, preventing the B cell from being activated. The B cells are not permanently inactivated, and when transferred to an antigen-free environment readily respond to immunogenic antigen doses.

Epitope density also influences immunogenicity and tolerogenicity. In general, increasing epitope density results in a decrease in immunogenicity and an increase in tolerogenicity.

Antigen dose

Many T-dependent antigens are immunogenic only within a certain dose range. Antigen concentrations above or below this range may be tolerogenic. These situations are called, respectively, *high zone tolerance* and *low zone tolerance*. Low and high zone tolerance appear to occur via different mechanisms. Low zone tolerance is due to inactivation of T_h cells, while high zone tolerance results from unresponsiveness of both T and B cells. It seems that mature T cells are more easily tolerized than mature B cells. T-independent antigens do not have the property of low and high zone tolerogenicity. They are only tolerogenic at high doses.

Mechanisms of tolerance

Since early experiments showed that tolerance was most easily induced in newborns, it has been proposed that self-nonself discrimination is programmed during intrauterine and early postnatal development. Clones reacting with self components were deleted from the system (*clonal deletion*). This concept was long accepted due to an inability to demonstrate the presence of precursors of B or T cells specific for self antigens.

Later, it was observed that immune responses following vaccination or during natural infections were sometimes accompanied by production of *rheumatoid factors*. These are autoantibodies specific for the F_c of IgG. Following the advent of hybridoma technology, autoreactive B or T lymphocytes could be immortalized. Analysis of hybridomas obtained from newborn animals, or human umbilical cord blood, showed that as many as 30% produce self-reactive antibodies. A significant fraction of healthy adults may also produce small quantities of self-reactive antibodies (natural autoantibodies). Clearly, self-specific clones are a normal component of the repertoire and are not deleted from the immune system. When these clones are activated and produce large amounts of autoantibodies, pathology may result (see Chapter 10).

It has also been observed that naive and antigen-tolerant animals have equal numbers of specific antigen-binding cells in their spleens. Thus, an alternative to the theory of clonal deletion is the theory of *clonal anergy* which explains B cell tolerance as a persistent inability of clones to respond to an immunogenic stimulus after having received a tolerogenic stimulus.

Anergy may be intrinsic to the B cell itself, or may be imposed on it from without, e.g., by active suppressor T cells or inactive helper T cells. Suppressor cells may exert their effects on B cells or on T_h cells. The difference between intrinsic clonal anergy and suppression (and lack of help) is that an anergic cell is incapable of antigen-specific activation. A cell which is not anergic, but is being actively suppressed, will be stimulated by appropriate signals when inhibitor cells and/or factors are removed from its environment. Both anergy and suppression may account for many aspects of natural and experimentally-induced B cell tolerance.

While clonal deletion does not appear to explain B cell tolerance, it may be involved in many aspects of T cell tolerance, particularly tolerance to self. This will be discussed further below.

B cell tolerance

Reagents which cross-link surface Ig (multivalent antigen or anti-Ig antibodies) may deliver a tolerizing signal. Immature B cells bearing only surface IgM are easily tolerized, while mature B cells bearing IgM and IgD are resistant to tolerization. Immature B cells and newly stimulated memory cells are dividing while mature B cells are in the G_0 (resting) phase of the cell cycle. Studies of immature B lymphoma cell lines showed that their proliferation could be halted early in the G_1 phase of the cell cycle after exposure to anti-Ig antibodies. These observations suggest that susceptibility to tolerogenic signals depends on the cell cycle, cells in early G_1 being vulnerable. This concept is supported by the observation that inhibitors of cell division affecting the G_1 and S phases (blockers of ATP generation, nucleic acid synthesis, and methyltransferases) interfered with tolerization, while colchicine, an inhibitor of mitosis, did not.

Reagents interacting with B cell surface Ig may modulate the activity of differentiated antibody-producing cells, as well as immature or resting B cells. Plasma cells and antibody-secreting lymphoma cell lines are frequently susceptible to negative signals mediated by their surface receptors. The amount of antibody secreted diminishes, and neoplastic cells may slow their rate of division. In a few instances, complete remissions of B cell lymphomas have been induced by treatment with anti-idiotypic antibodies specific for the lymphoma Ig. This mechanism may explain some aspects of self-tolerance. Newly formed B cells expressing surface Ig, but not having performed their final division, may be tolerized if they encounter antigen in sufficient quantities to cause Ig cross-linking.

As described above, B cells in the early G_1 phase of the cell cycle are the most susceptible to tolerization by Ig receptor cross-linking. Just as immunogenicity depends on the affinity of the interaction of antigen and antibody, so does tolerogenicity. As the affinity of the B cell receptor for antigen increases, so does the potential for tolerization, i.e., a B cell with higher affinity receptors is tolerized by lower antigen concentrations.

Antigen-antibody complexes may be tolerogenic. This effect appears to require simultaneous occupancy (cross-linking?) of surface Ig and Fc receptors since $F(ab)'_2$-antigen complexes under the same conditions are not tolerogens.

A stimulus that would normally be tolerogenic may be rendered immunogenic if T cell help is added to the equation. A particular hapten-carrier conjugate administered in a (hapten-specific) tolerogenic dose will become (hapten-specific) immunogenic if the animal is first immunized to the carrier. The same is true for antigen-antibody complexes. These are no longer tolerogenic if the animal is first primed against the antibodies in the complexes.

Some researchers have begun to examine the responses of tolerant cells to cytokines. Tolerant murine B cells still respond to IL-4 by increasing MHC class II antigen surface density by the same amount as normal B cells. Crude mixtures of T cell cytokines prevent tolerance induction if given concomitantly with the tolerizing stimulus. IL-4 and IL-5 are necessary, but not sufficient to produce this effect.

There is much evidence that the anergic state is reversible in some instances. Transfer of B cells from a tolerant animal to an irradiated host or tissue culture may allow return to normal function. This may be explained by release of B cells from suppression by T_s, but it seems clear that in some instances there is an actual change in the function of the B cells themselves. The importance of the persistence of tolerogen for maintaining tolerance in many instances suggests that constant occupancy of Ig receptors may be necessary for continuation of the anergic state.

T cell tolerance

As with B cells, the mechanisms of T cell tolerance remain predominantly speculative. While clonal anergy and suppression probably play a role in both B and T cell tolerance, clonal deletion appears to be important in only T cell tolerance, especially tolerance to self antigens.

One of the most striking aspects of T cell activation is that these cells are not stimulated by self MHC molecules alone, but are activated when foreign antigen fragments associate with them. With respect to antigens, T cell tolerance is frequently categorized as tolerance to self MHC, to self antigens other than MHC, or to foreign antigens.

Most researchers believe that tolerance to self MHC is the result of a clonal deletion process. Clones reacting to self either die or are destroyed before they leave the thymus. A deletion model of selection is quite plausible considering that $>90\%$ of the pre-T cells entering the thymus do not leave. Is this passive cell death, or active cell killing? Some have proposed that the majority of T cells die in the thymus by a process of *apoptosis*. This is a pre-programmed series of intracellular metabolic events resulting in the degradation of genomic DNA and eventual cell death.

In one model of thymic T cell development, all entering pre-T cells have started along the pathway of apoptosis. Pre-T cells may then be protected from death by appropriate cellular interactions. This type of positive selection is necessary to ensure that mature T cells will be functional within the context of self MHC antigens. That is, a minimal affinity of interaction of T cell receptors with self MHC is necessary so that mature cells will be effectively stimulated by antigen complexed with them. As described in Chapter 7, it has been suggested that peptides which mimic foreign antigens are generated in the thymus via mistranscription of DNA and mistranslation of mRNA. This would (possibly) permit the positive selection of the entire T cell repertoire. Tolerance to self MHC then requires negative selection of those T cells having too high an affinity for self MHC. These cells are destroyed. Two different cell populations are thought to mediate these selective processes. Cortical thymic epithelial cells carry out positive selection, while macrophages and dendritic cells at the corticomedullary junction perform negative selection.

Another hypothetical mechanism of T cell tolerance states that the principal interaction determining T cell tolerance to self MHC is intracellular rather than extracellular. Recall that T cells (as do all cells) bear MHC class I molecules. If a T cell creates an antigen receptor capable of interacting with its own MHC class I antigens, the endoplasmic reticulum may be irreversibly congested with complexes formed by these proteins. This condition is presumably incompatible with further development of the T cell. According to this model, tolerance to MHC class II then arises by simple extension of tolerance to class I, i.e., class II antigens are sufficiently similar to class I such that any cell capable of recognizing class II will recognize class I and be eliminated.

Mechanisms of active suppression, or an intrinsic anergic state have also been implicated in several experimental systems of T cell tolerance to self. T cell anergy may be explained in the following way: a single signal delivered by the T cell's TCR-CD3 complex is tolerogenic, the same signal together with additional signals (cytokines and/or other cell-surface interactions) provided by the antigen-presenting cell constitutes an immunogenic stimulus. A tolerogenic interaction might then be provided by any cell expressing MHC class II molecules which is incapable of providing additional stimulatory signals. Neonatal spleen cells have these properties.

Recently, a type of $CD8^+$ T cell called a *veto cell* has been implicated in several systems of T cell tolerance to self MHC class I, minor histocompatibility antigens, as well as foreign antigens. These cells act in a MHC haplotype-specific manner, indicating that mutual cellular recognition is important for their activity. Veto cells are distinct from suppressor cells in that they actually reduce the numbers of cytotoxic cell precursors, as well as rendering them permanently incapable of being activated. It does not appear that veto cells actually destroy precursors, so the mechanism of this decrease is unclear.

As with B cells, induction of the anergic stage in T cells requires complex intracellular metabolic processes. This induction is independent of the phosphatidylinositol pathway leading to cellular activation. Tolerized cells fail to secrete IL-2.

Breaking tolerance

Although tolerance may be long-lasting, it is frequently impermanent. An animal may spontaneously escape from tolerance, or it may be deliberately ended. As mentioned above, persistence of antigen appears to be important for maintenance of the tolerant state in many instances. In other situations tolerance to self (or foreign) antigens may be broken by inoculation of a different cross-reactive antigen. For example, rabbits injected with rat thyroglobulin (TG) produce anti-rat TG antibodies reacting with rabbit thyroglobulin. Tolerance may also be broken by administration of mitogens or adjuvants causing nonspecific cellular proliferation.

In Chapter 10 we will examine in more detail some ways in which our immune systems might lose tolerance to self and generate harmful autoimmune responses.

Immunosuppression

Occasionally, our therapeutic goal is to suppress the activity of the immune system. Harmful immune responses occur in autoimmune diseases and some parasitic infections. In modern medical practive however, this situation arises most frequently in organ or tissue transplantation. The graft contains many antigenic determinants to which the recipient is not tolerant. Thus, the graft is rejected. Alternatively, immunocompetent cells in bone marrow grafts given to immunocompromised hosts are stimulated and begin to "reject" the host (graft versus host disease, see Chapter 8). In both of these instances, one wishes to artificially suppress or induce tolerance within the immune system. Several techniques have been employed with varying success, and new methods are being developed rapidly.

Drugs

High-dose corticosteroids are very effective immunosuppressants, but are associated with many adverse systemic effects. Many cytotoxic antineoplastic agents are also immunosuppressive due to their interference with hemopoiesis. Although these drugs are effective in immunosuppression, their severe side effects make them unattractive for this specific purpose.

Cyclosporine, an antibiotic produced by soil fungi, is a potent immunosuppressant. This drug has increased mean survival of heart transplant recipients by 40%, and kidney transplant recipients by 30%. This drug exerts its effects non-specifically by blocking either production of, or the activity of IL-2. This appears to selectively inhibit induction of $T_{h/i}$ cells, but not T_s cells. Cyclosporine also potentiates the effects of other immunosuppressants such as anti-lymphocyte antibodies (see below). Specific suppression for donor MHC antigens may be achieved by injecting the host with extracts of donor MHC simultaneously with cyclosporine and corticosteroids.

Recently, another fungus-derived antibiotic known as FK-506 has been used in immunosuppressive regimens to prolong graft survival. FK-506 is more potent than cyclosporine by approximately a factor of 10 (by mass), however, it is highly toxic. Since FK-506 synergizes with cyclosporine, these drugs may be used together and with other immunosuppressive modalities, perhaps, in regimens that will have reduced toxicity.

Anti-lymphocyte antibodies

Anti-lymphocyte antibodies may induce complement lysis or antibody-dependent cell-mediated cytotoxicity against cells to which they bind. These reagents, both monoclonal and polyclonal, are used in two different ways: *ex vivo* to eliminate T cells from bone marrow grafts to prevent graft-versus-host disease, or *in vivo* in the host to eliminate peripheral T cells capable of reacting to a grafted organ. Antibodies against CD4, CD8, and the T cell receptor have all been shown to be very immunosuppressive in mice and rats. Additional antigens being evaluated as targets for immunosuppressive antibodies include CD2, CD5, CD7, CD11/18, and CD25.

Antibodies specific for CD3 have been used to suppress graft rejection in kidney transplants. Anti-CD3 antibodies obtained from mice were administered over a one month period following transplantation. An immune response to the mouse antibodies was prevented with the cytotoxic and immunosuppressive drug azathioprene. Patients treated with anti-CD3 had improved graft survival and a lower requirement for maintenance immunosuppressive therapy than did patients treated only with steroids. The results with anti-CD3 were comparable to those obtainable with cyclosporine.

Recombinant DNA technology has been used to produce chimeric antibodies, i.e., with V regions from one species and C regions from another. Immune response to antibodies from other species are primarily directed against C regions. Thus, the use of chimeric antibodies with the requisite specificity (e.g., anti-human CD3 from mice) in association with human C regions could eliminate the need for additional immunosuppression in protocols using anti-lymphocyte antibodies.

Radiation

X-irradiation together with antigen administration induces tolerance in a manner similar to cyclosporine. Radiation also synergizes with anti-lymphocyte antibodies, and may be used with donor MHC immunization.

Pregnancy

A physiologic form of immunosuppression occurs during pregnancy. A selective decrease in cell-mediated immunity is observed in gravid females, while humoral immunity is intact. This phenomenon may be necessary for the mother not to reject the fetus immunologically. Fetal tissues are much like an allograft in that they contain alien histocompatibility antigens derived from the father. The levels of a multitude of hormones and serum proteins undergo great changes during pregnancy. Several of these have been shown to be immunosuppressive *in vitro*. These include estrogen, progesterone, cortisol, α fetoprotein, and uromodulin.

The immunologic distinction of self and nonself is not a straightforward one, and probably follows rules utterly incompatible with the anthropomorphic patterns of thinking in which that distinction is most frequently expressed ("this is me, this is not me"). While the immune system's view of the universe remains less than clear to us, we may at least say that usually it treats us well, and at other times not so well. The latter situation is the focus of the next chapter.

SOURCES AND SUGGESTED ADDITIONAL READING

Ada, G. L. & Parish, C. R. (1968) Low zone tolerance to bacterial flagellin in adult rats: a possible role for antigen localized in lymphoid follicles. *Proc. Natl. Acad. Sci. USA*, **61**:556–561.

Battisto, J. R. & Miller, J. (1962) Immunological unresponsiveness produced in adult guinea pigs by parenteral introduction of minute quantities of hapten or protein antigen. *Proc. Soc. Exp. Biol. Med.*, **111**:111–115.

Billingham, R. E. & Silvers, W. K. (1961) Quantitative studies on the ability of cells of different origins to induce tolerance of skin homografts and cause runt disease in neonatal mice. *J. Exp. Zool.*, **146**:113–129.

Burnet, F. M. & Fenner, F. (1949) *The production of antibodies*, 2nd ed., McMillan & Co., Melbourne.

Chase, M. W. (1946) Inhibition of experimental drug allergy by prior feeding of the sensitizing agent. *Proc. Soc. Exp. Biol. Med.*, **61**:257–259.

Debure, A., Chkoff, N., Chatenoud, L., Lacombe, M., Campos, H., Noël, L. H., Goldstein, G., Bach, J. F. & Kreis, H. (1988) One-month prophylactic use of OKT3 in cadaver kidney transplant recipients. *Transplantation*, **45**:546–553.

Horiuchi, A. & Waksman, B. H. (1968) Role of the thymus in tolerance. VIII. Relative effectiveness of non-aggregated and heat-aggregated bovine γ globulin, injected directly into lymphoid organs of normal rats, in suppressing immune responsiveness. *J. Immunol.*, **101**:1322–1332.

Howard, J. G., Christie, G. H., Jacob, M. J. & Elson, J. (1970) Studies on immunological paralysis. III. Recirculation and antibody-neutralizing activity of ^{14}C-labelled type III pneumococcal polysaccharide in paralysed mice. *Clin. Exp. Immunol.*, **7**:583–596.

Janeway, C. A., Jr. & Sela, M. (1967) Synthetic antigens composed exclusively of L- or D-amino acids. I. Effect of optical configuration on the immunogenicity of synthetic polypeptides in mice. *Immunology*, **13**:29–38.

Jenkins, M. K., Pardoll, D. M., Mizuguchi, J., Quill, H. & Schwartz, R. H. (1987) T-cell unresponsiveness *in vivo* and *in vitro*: fine specificity of induction and molecular characterization of the unresponsive state. *Immunol. Rev.*, **95**:113–135.

Kahan, B. D., Didlake, R., Kim, E. E., Yoshimura, N., Kondo, E. & Stepkowski, S. (1988) Important role of cyclosporine for the induction of immunologic tolerance in adult hosts. *Transplant. Proc.*, **20** (Suppl. 3):23–35.

Klinman, N. R., Riley, R. L., Morrow, P. R., Jemmerson, R. R. & Teale, J. M. (1985) Tolerance and B cell repertoire establishment. *Federation Proc.*, **44**:2488–2492.

MacDonald, H. R., Pedrazzini, T., Schneider, R., Louis, J. A., Zinkernagel, R. M. & Hengartner, H. (1988) Intrathymic elimination of Mls^a-reactive ($V_\beta 6^+$) cells during neonatal tolerance induction to Mls^a-encoded antigens. *J. Exp. Med.*, **167**:2005–2010.

Maki, T. (1988) Suppressor cells and mediators as immunoregulators. *Transplant. Proc.*, **20**:1204–1206.

Miller, J. F. A. P. & Watson, J. D. (1988) Intracellular recognition events eliminate self-reactive T cells. *Scand. J. Immunol.*, **28**:389–395.

Mitchison, N. A. (1965) Induction of immunological paralysis in two zones of dosage. *Proc. R. Soc. Lond. (Ser. B)*, **161**:275–292.

Moreno, C., Hale, C. & Ivanyi, L. (1977) The mitogenic, immunogenic and tolerogenic properties of dextrans and levans. Lack of correlation according to differences of molecular structure and size. *Immunology*, **33**:261–267.

Morrison, S. L., Johnson, M. J., Herzenberg, L. A. & Oi, V. T. (1984) Chimeric human antibody molecules: mouse antigen-binding domains with human constant region domains. *Proc. Natl. Acad. Sci. USA*, **81**:6851–6855.

Nossal, G. J. V. (1983) Cellular mechanisms of immunologic tolerance. *Ann. Rev. Immunol.*, **1**:33–62.

Owen, R. D. (1957) Erythrocyte antigens and tolerance phenomena. *Proc. R. Soc. Lond. (Ser. B)*, **146**:8–18.

Scott, D. W., Chace, J. H., Warner, G. L., O'Garra, A., Klaus, G. G. & Quill, H. (1987) Role of T cell-derived lymphokines in two models of B cell tolerance. *Immunol. Rev.*, **99**:153–171.

Scott, D. W. & Klinman, N. R. (1987) Is tolerance the result of engaging surface Ig of B cells in cycle? *Immunol. Today*, **8**:105–106.

Sprent, J., Lo, D., Gao, E.-K. & Ron, Y. (1988) T cell selection in the thymus. *Immunol. Rev.*, **101**:173–190.

Thomson, A. W. (1989) FK-506—How much potential? *Immunol. Today*, **10**:6–9.

Waldmann, H. (1988) Immunosuppression with monoclonal antibodies: some speculations about tolerance in the context of tissue grafting. *Transplant. Proc.*, **20** (Suppl. 8):46–52.

Weigle, W. O. (1971) Recent observations and concepts in immunological unresponsiveness and autoimmunity. *Clin. Exp. Immunol.*, **9**:437–447.

Weinberg, E. D. (1987) Pregnancy-associated immune suppression: risks and mechanisms. *Microb. Pathol.*, **3**:393–397.

Wells, H. G. & Osborne, T. B. (1911) The biological reactions of the vegetable proteins. I. Anaphylaxis. *J. Infect. Dis.*, **8**:66–125.

Chapter 10

Immunopathology

"To err is human", and our immune systems are neither more nor less frail than other parts of our bodies. Immune responses to invading microbial antigens, and others contacted in the environment, may produce harmful effects. The immune system may even lose tolerance to self and generate humoral and cell-mediated responses against our own macromolecular constituents. As the word "pathology" designates any deleterious process within the tissues and systems of an organism, so its derivative *immunopathology* designates those processes arising from immune system dysfunction, or harmful "side effects" of its physiologic functions.

The various immune phenomena leading to tissue injury have been called *hypersensitivity*. These immune reactions may be classified into five types:

Type I —Immediate hypersensitivity (anaphylaxis).
Type II —Antibody-dependent cytotoxicity.
Type III—Immune complex reactions.
Type IV—Delayed hypersensitivity.
Type V —Stimulatory reactions.

Type I reactions occur in immune responses to exogenous antigens. Types II, III, and IV may occur in responses to exogenous or endogenous (self) antigens. Type V reactions are a subset of the diseases resulting from production of anti-self-antigen antibodies. Below, we will describe the mechanisms underlying types I–III, and discuss them in the context of exogenous antigens. Then we will describe together the diseases resulting from these reactions in the context of endogenous antigens (autoimmune diseases). Type IV reactions (delayed hypersensitivity) have been described in Chapter 7. DTH has also been implicated in the etiology of several autoimmune processes.

TYPE I. IMMEDIATE HYPERSENSITIVITY

Some early studies of hypersensitivity examined the effects of a sea anemone toxin injected into dogs. When dogs who survived an initial large dose were injected weeks later with very small amounts of toxin, they died quickly. Thus, the first administration of toxin resulted not in a "resistance" to the toxin, but a "hypersensitivity." The term *anaphylaxis* was coined (from the Greek *ana*—away from, and *phylaxis*—protection) to describe this phenomenon. Later it was learned that anaphylaxis is mediated by the immune system.

Antigens eliciting anaphylactic reactions are called *allergens*, and they are responsible for our *allergies*. These substances may be divided into two broad categories, naturally-occurring allergens, for example, animal danders, plant pollens, and some foods; and synthetic molecules such as acetylsalicyclic acid (aspirin), and other drugs. Allergens have no general chemical or structural distinguishing features, but they are all similar in that they may elicit predominantly IgE antibody responses in certain individuals, or at certain times in one's life. IgE antibodies are also called *homocytotropic* or *reaginic antibodies*. Predominance of the IgE isotype appears to be under T cell control. The IgE molecule is the *sine qua non* of immediate hypersensitivity.

As in other primary immune responses, the first encounter with allergen results in IgM synthesis, with other isotypes (IgE) only appearing late. Basophils and mast cells possess high-affinity receptors for the F_c of IgE ($Fc_{\varepsilon}RI$) which adsorb it from the circulation. When the circulation. When the allergen is encountered again, it may react with cell-bound IgE. If some fraction of the receptor-IgE complexes are cross-linked by allergen, *degranulation* of the mast cell ensues (Figure 10.1). The granules contain biologically active substances called *mediators* of the hypersensitivity reaction (Table 10.I). In addition to releasing preformed mediators from granules, mast cells begin to synthesize other mediators following IgE F_c-receptor cross-linking. The reaction may begin in minutes, hence, the term *immediate hypersensitivity*. Immediate hypersensitivity reactions may be local (a nuisance) or systemic (often fatal).

Local immediate hypersensitivity

The passive transfer of cutaneous immediate hypersensitivity is known as the *Prausnitz-Küstner reaction*, after the physicians who first demonstrated this phenomenon. Küstner, who was allergic to fish,

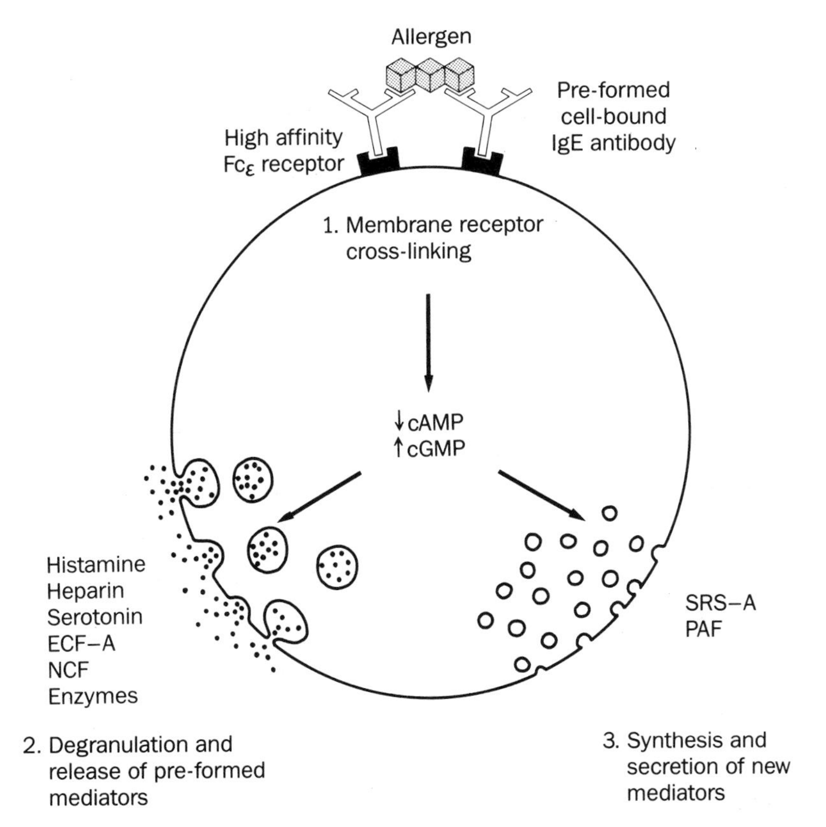

Figure 10.1. *Immediate hypersensitivity (anaphylaxis).* Mast cells and basophils possess high affinity receptors for the Fc of IgE ($Fc_{\varepsilon}RI$). A portion of the IgE produced in immune responses binds to these receptors. When antigen (allergen) subsequently contacts these cells, it may cross-link the receptors by simultaneous binding to two or more IgE combining sites. This signal is transduced biochemically via cyclic nucleotides, and stimulates two processes: release of mediators pre-formed and packaged in cytoplasmic granules, and *de novo* synthesis and secretion of additional mediators.

injected a small quantity of his serum subcutaneously into Prausnitz. When a fish extract was injected into the same site in Prausnitz, a wheal and flare reaction ensued, identical with one seen in Küstner. This experiment demonstrated that a serum factor, later identified as IgE antibodies, was responsible for hypersensitivity.

Local immediate hypersensitivity is limited to the area of encounter with the allergen, a patch of skin, the nasal mucosa, etc. Skin reactions are primarily erythema, pruritus (itching) and urticaria (hives). These phenomena are often called a *wheal and flare* reaction. Any hay fever

Table 10.I. MEDIATORS OF HYPERSENSITIVITY AND INFLAMMATION

Mediator	Actions
Histamine	Increased vascular permeability Vascular smooth muscle relaxation Non-vascular smooth muscle contraction
Serotonin	Increased vascular permeability Smooth muscle contraction
Complement fragments (C3a and C5a)	Mast cell/basophil degranulation PMN chemotaxis and degranulation Increased vascular permeability Smooth muscle contraction
Arachidonic acid derivatives	
prostaglandins	Inhibit: leukocyte chemotaxis and degranulation, mast cell/basophil degranulation platelet aggregation
leukotrienes	Smooth muscle contraction PMN chemotaxis and degranulation
thromboxanes	Smooth muscle contraction Platelet aggregation
Eosinophil chemotactic factor of anaphylaxis	Eosinophil chemotaxis
Platelet activating factor	Platelet aggregation
Kinins	Increased vascular permeability Smooth muscle contraction PMN chemotaxis and degranulation
Heparin	Anticoagulant

sufferer can describe the irritation and congestion resulting when pollen grains deposit on the nasal mucosa. The substance primarily responsible for these cutaneous and mucosal phenomena is *histamine* released from the mast cell granules, thus, antihistamines may provide some relief. Histologically, small blood vessels dilate and fluid accumulates in extravascular spaces. Late in the reaction, eosinophilia may be evident to varying degrees.

As a rule, local reactions last as long as contact with the allergen persists. If encounter with allergen is transient, the reaction usually resolves within 24 hours. If the irritating substance persists, however, suffering may be prolonged.

The most common diagnostic test for allergy is a *prick test*. Drops of solutions or suspensions of common allergens (animal danders, pollens, venoms, foods, drugs, etc.) are placed on the skin, usually the forearm. A fine needle is then touched to the skin under a drop, just enough to break the epidermis without going through the skin. A wheal and flare reaction developing in 10–20 minutes indicates allergy to that substance.

Systemic anaphylaxis

When allergen enters the circulation of a sensitized individual, a systemic anaphylactic reaction ensues. The severity of the reaction depends on the allergen dose and the amount of specific IgE bound to mast cells. In many instances, systemic anaphylaxis is rapidly fatal. Death usually results from severe bronchospasm and laryngeal edema coupled with rapidly falling blood pressure and hemoconcentration. Vomiting, diarrhea, urinary bladder spasm, and cerebral edema may also occur. Individuals progressing to hypotonic shock rarely survive without heroic measures. If given early enough, intramuscular or intravenous injection of epinephrine may prevent most of the life-threatening manifestations of systemic anaphylaxis. Epinephrine inhibits degranulation of mast cells and basophils.

Mediators of immediate hypersensitivity

As mentioned above, mediators of immediate hypersensitivity are either preformed (stored in mast cell or basophil granules), or newly synthesized following triggering of the cell. Biologically active compounds mediating hypersensitivity are summarized in Table 10.I. Throughout this discussion, we shall not distinguish between mast cells and basophils. Although these cell types have many important attributes in common, the reader should be aware that differences between them exist. In addition, there is even heterogeneity within each population. These different cell types may play subtly different roles in various hypersensitivity phenomena.

Preformed mediators

Histamine is stored in mast cell granules in a proteoglycan complex. Derived by decarboxylation of histidine, histamine is a CNS neurotransmitter; it is stored in many sites in the body; and has many pharmacological effects. Important effects in anaphylaxis are the relaxation of vascular smooth muscle, contraction of non-vascular (bronchiolar, urinary bladder, GI tract) smooth muscle, and increased secretion by glandular tissues.

The tryptophan derivative *serotonin*, whose chemical name is 5-hydroxytryptamine (5-HT), is also found in the brain as a neurotransmitter. Additional sites of storage are gastrointestinal chromaffin cells, basophils and platelets. This hormone causes smooth muscle contraction and increases capillary permeability.

The mediator *eosinophil chemotactic factor of anaphylaxis* (*ECF-A*) is stored in granules as a mixture of four tetrapeptides. These substances induce the accumulation of eosinophils at the site of release. Eosinophilia is a marker of *atopic diseases*. Individuals prone to *atopy* tend to produce IgE responses more readily than others and, hence, suffer much symptomatology of immediate hypersensitivity. One of the major atopic diseases is bronchial asthma, in which eosinophilia of respiratory mucosae may be pronounced. *Neutrophil chemotactic factor* is a high-molecular-weight protein ($M_r > 160{,}000$). As its name implies, it is chemotactic for neutrophils.

The sulfated proteoglycan *heparin* is best known for its anticoagulant action. Although it undoubtedly exerts this effect to some extent when released from mast cell granules, heparin may also have other roles in immediate hypersensitivity. Evidence indicates that it may moderate the reaction in part by inhibiting the generation of complement fragments or their effective association.

Other mast cell/basophil granule contents

In addition to the above-mentioned substances, mast cell granules contain a variety of enzymes. Some of these have activities similar to (but are distinct from) serum proteases such as trypsin, chymotrypsin, and kallikrein. These proteases may cleave complement components yielding the anaphylatoxins C3a and C5a, and kininogen yielding *kinins*. C3a induces mast cell and basophil degranulation, smooth muscle contraction (which may be secondarily mediated by histamine), and an increase in mucus secretion. C5a has the same effects and is also chemotactic for phagocytic cells and increases vascular permeability. The kinins are pharmacologically active peptides with effects similar to histamine.

Newly-synthesized mediators

Arachidonic acid gives rise to many pharmacologically diverse substances. Some of these products are classified into three main groups based on their chemical structure. They are the *prostaglandins*, *leukotrienes*, and *thromboxanes*. Activated mast cells and basophils synthesize several of these compounds. What has long been known as *slow-reacting substance of anaphylaxis* (*SRS-A*) is a mixture of leukotrienes C_4, D_4, and E_4 (LTC_4, LTD_4, LTE_4). Leukotriene B (*LTB*) and prostaglandin D_2 (PGD_2) are the other principal arachidonate derivatives secreted. The pharmacology of these substances is complex. Some important actions of the leukotrienes are the stimulation of phagocyte chemotaxis, and release of neutrophil lysosome contents. PGD_2 inhibits platelet aggregation. Another prostaglndin, PGE_2, may function in suppressing the immediate hypersensitivity reaction. Neutrophils and macrophages synthesize large amounts of PGE_2 which inhibits mediator release from mast cells and basophils.

Platelet activating factor (*PAF*) is a group of phosphorylcholine derivatives. Biological effects include platelet and neutrophil aggregation, and increased vascular permeability. PAF may also cause contraction of pulmonary vascular and bronchial smooth muscle. Whether this is a direct effect or is mediated through prostaglandins is not clear. PAF also produces wheal and flare reactions, although the mechanism is unknown.

A diagram summarizing the events in immediate hypersensitivity reactions is presented in Figure 10.1.

Treatment of immediate hypersensitivity

One of the most common allergies is to plant pollens, most notably plants of the genus inappropriately named *Ambrosia*, or the ragweeds. For most sufferers, this is a more or less short-lived annual nuisance that can be at least partially alleviated with antihistamines. Some individuals have particularly severe allergies to pollens, and their reactions are literally prostrating. Allergies to bee or wasp venoms may be especially dangerous. Every chance meeting of a sensitive individual with one of these stinging insects could result in death if appropriate treatment (epinephrine) is not readily available.

In these more serious allergies, it is desirable to interrupt the mechanisms of immediate hypersensitivity. In some instances a process of *desensitization* may be effective. Initially, small doses of allergen are administered, usually intramuscularly. Over time (months) the dosage is increased. The aim is to increase the synthesis of IgG in favor of IgE. Circulating IgG reacts with the allergen before it contacts cell-bound IgE.

TYPE II. ANTIBODY-DEPENDENT CYTOTOXICITY

The Type II reactions result from antibody binding to cell-surface antigens. Lysis occurs by either of two mechanisms: activation of the complement system (see Chapter 4); or by antibody-dependent cell-mediated cytotoxicity (ADCC, see Chapter 7). Recall that IgG and IgM are effective complement activators, while ADCC is mediated by IgG only.

Harmful effects of complement activation arise by direct lysis of cells to which antibody has bound, and by recruitment of other leukocytes by complement products. C5a is chemotactic for phagocytes and both C3a and C5a stimulate release of their lysosomal contents, including proteases which may generate kinins, and others with directly cytotoxic properties. Mast cell and platelet degranulation (releasing vasoactive amines) are additional indirect effects of complement activation. These inflammatory processes lead to destruction of tissue to which antibody has bound. Indirect inflammatory effects of complement activation are diagrammed in Figure 10.2.

Red blood cells possess many different surface antigens, over 400 have been identified to date. Some are shared with many body tissues, some are erythrocyte-specific. The ABO group of polysaccharide antigens is one of the most well-studied. Approximately six months after birth, antibodies reacting with ABO antigens *not* found in the body began to appear in the blood. These antibodies are called *naturally occurring isohemagglutinins* (antibodies agglutinating red blood cells). The stimulus initiating isohemagglutinin production is unknown, but some suggest that these antibodies are produced after contact with bacterial polysaccharides and cross-react with ABO antigens. If blood is administered to a person with isohemagglutinins specific for ABO

▶

Figure 10.2. *Complement activation and inflammation.* Both cell-bound antibodies and immune complexes may initiate the classical pathway of complement activation. The anaphylatoxins C3a and C5a directly induce mast cell degranulation, C5a also recruits neutrophils to the site of activation. These cells release biologically active substances (mediators) as well as degradative enzymes. The former modulate the activity of leukocytes and the functions of vascular smooth muscle and endothelium, while the latter cause tissue injury and also activate the kinin system which synergizes with other mediators. Increases in vascular permeability and smooth muscle relaxation lead to localized edema. Tissue destruction increases complement activation via the alternate pathway, and initiates the clotting cascade. Thrombosis is also promoted by mediator-induced platelet aggregation.

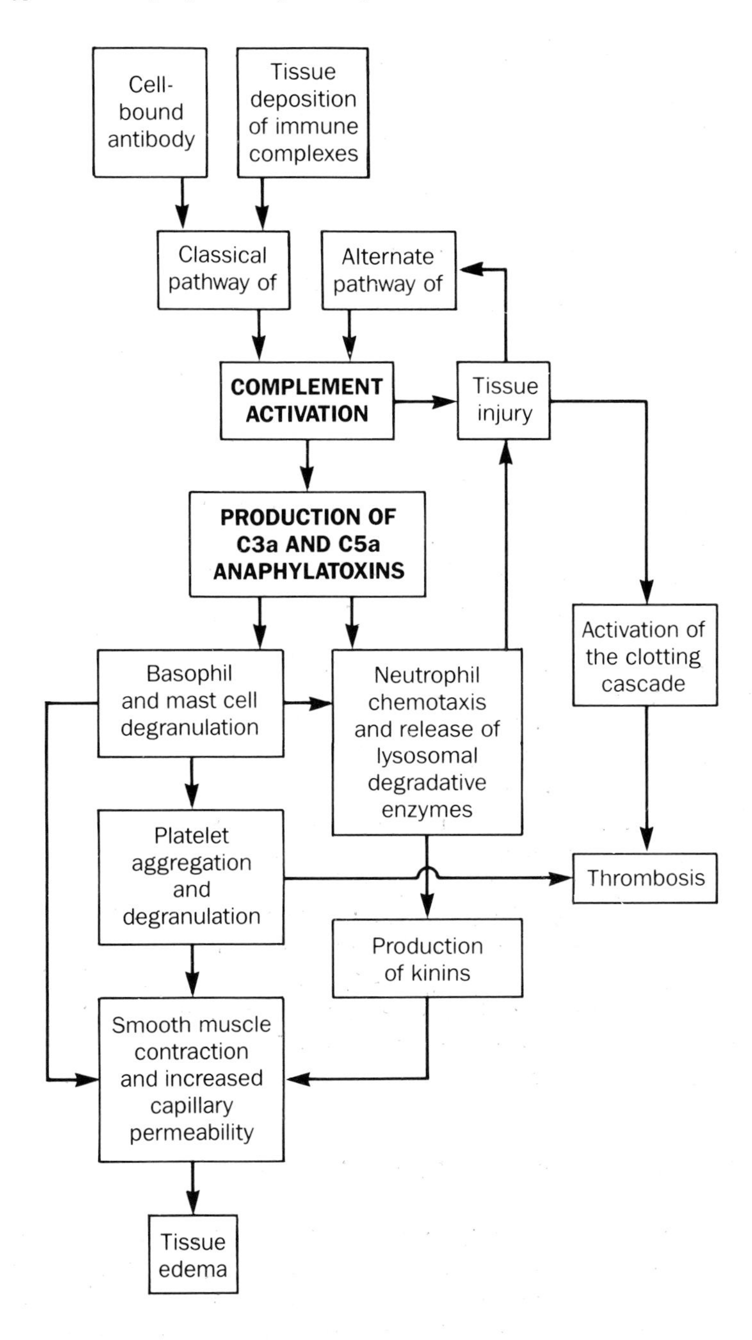
Cell-bound antibody
Tissue deposition of immune complexes
Classical pathway of
Alternate pathway of
COMPLEMENT ACTIVATION
Tissue injury
PRODUCTION OF C3a AND C5a ANAPHYLATOXINS
Activation of the clotting cascade
Basophil and mast cell degranulation
Neutrophil chemotaxis and release of lysosomal degradative enzymes
Platelet aggregation and degranulation
Thrombosis
Production of kinins
Smooth muscle contraction and increased capillary permeability
Tissue edema

antigens in the transfused blood, a *transfusion reaction* ensues. Isohemagglutinins coat the foreign red cells which are then susceptible to complement lysis, opsonization and phagocytosis (see Chapter 4). The destruction of transfused cells combined with widespread activation of the complement and kinin systems may be fatal. Signs and symptoms of the transfusion reaction include fever, chills, hypotension, nausea, vomiting, hemoglobinemia, hemoglobinuria, and renal failure.

Hemolytic disease of the newborn or *erythroblastosis fetalis* occurs when maternal IgG antibodies binding fetal red cell antigens cross the placenta into the fetal circulation. This may occur in ABO incompatible mothers and babies, but is more common in the case of *Rh* (*rhesus*) incompatibility. Rh is another erythrocyte antigen system. Small numbers of fetal red cells are not uncommon in maternal blood during gestation. In addition, at birth, fairly large numbers of the neonate's red cells may enter the maternal circulation where they induce an antibody response to any antigens not shared with the mother. This sensitization may yield a vigorous secondary response if the antigen is encountered again in a subsequent pregnancy. Transplacental transfer of IgG antibodies into the fetal circulation results in fetal red cell destruction and anemia which may be fatal. When Rh incompatibility has been determined, maternal sensitization following delivery may be prevented if she is passively immunized with *Rh immune globulin.* These antibodies bind the Rh antigen on fetal red cells in her circulation, and prevent it from stimulating an immune response.

Type II reactions occur in *Coombs' positive hemolytic anemia* (see Chapter 5) which manifests in some autoimmune diseases and in reactions to certain drugs. Compounds complexing with erythrocyte surface antigens may stimulate an immune response against the resulting "neoantigen." These antibodies behave as isohemagglutinins. Some of the drugs capable of initiating this type of reaction are methyldopa, stibophen, penicillin, and mefenamic acid. The same situation may occur with respect to platelets or neutrophils resulting in thrombocytopenia or neutropenia.

TYPE III. IMMUNE COMPLEX REACTIONS

As described in Chapter 5, antibodies are multivalent and may form complexes containing several molecules of antibody and antigen. These immune complexes may form in the blood and be deposited in small blood vessels, or they may form extravascularly when antibodies diffuse out and encounter antigen. Very large complexes are rapidly cleared by the mononuclear phagocyte system (MPS), small completes persist in the circulation longer, also increasing opportunity for clearance by the MPS. Only complexes of an intermediate size deposit rapidly in tissues and initiate harmful reactions. Once again, complement activation plays a central role in the inflammatory reaction. The processes involved are identical to those depicted in Figure 10.2 for the type II reaction. The difference is that activation in the type II reaction is by cell-bound antibody, while in the type III reaction, complement is activated by immune complexes. As with immediate hypersensitivity, immune complex reactions may be local or systemic.

Local immune complex disease: the Arthus reaction

Intradermal injection of antigen into a previously immunized animal generates immune complexes at the injection site. These complexes consist of preformed antibody and the injected antigen. Complement activation leads to neutrophil infiltration of the injection site and tissue destruction. The reaction is detectable 1–3 hours post-injection and reaches a peak after 4–10 hours. The injection site becomes edematous, erythematous, and may ulcerate. Histologically, one observes fibrinoid necrosis. Antibody and complement are detectable with immunostaining. A "passive" Arthus reaction may be induced by simultaneous intradermal injection of antigen and antibody.

The Arthus reaction is an experimental phenomenon induced in the skin. Several human diseases appear to result from similar processes. The synovial and cutaneous lesions of rheumatoid arthritis (RA) may be due to *antibody-antibody* complex deposition in these areas. *Rheumatoid factors* (*RFs*) are antibodies specific for the F_c of IgG produced in the same individual. RFs are found in the majority of adult patients with rheumatoid arthritis, and in about 50% of children with RA. However, rheumatoid factors are also found in varying amounts in individuals with no symptomatology. For this reason, some controversy exists concerning their pathogenicity.

Renal glomeruli are frequently the site of immune complex deposition. *Immune complex glomerulophritis is often* associated with autoimmune syndromes in which circulating immune complexes form (see below). The focal necrosis in blood vessel walls seen in *polyarteritis nodosa* may also be due to immune complex deposits. In some studies, as many as 40% of patients with this disease had immune complexes containing hepatitis B surface antigen in arterial lesions. Fungal and bacterial pulmonary infections may engender type III reactions resulting in *hypersensitivity pneumonitis.*

Systemic immune complex disease: serum sickness

People infected with *C. tetanii* may die from the effects of the tetanus toxin, if they have not been previously immunized with the toxoid. Tetanus antiserum, prepared in horses, neutralizes the toxin while the infection is brought under control. Approximately 10 days following injection of antiserum, some patients develop a diffuse skin rash, arthralgia, fever, splenomegaly, and proteinuria. Most individuals recover in a few days.

These symptoms arise when large amounts of a foreign protein are injected into the circulation. When specific antibody production begins, immune complexes form. These complexes leave the circulation and deposit in a variety of tissues. Widespread activation of complement and neutrophilic infiltration of multiple tissues generate the symptomatology of serum sickness. The process whereby immune complexes leave the circulation and deposit in tissues is not entirely clear, but there is evidence for involvement of a minor IgE component of the total specific antibody produced. This IgE binds to mast cells which release platelet activating factor (see above). Platelets aggregate and degranulate releasing vasoactive amines and causing localized increases in vascular permeability. These areas may be favored sites for the emigration of immune complexes into extravascular spaces.

AUTOIMMUNE DISEASES

Characteristic variations of macromolecular structure allow us to determine the species origin of many proteins and glycoproteins. For example, mice and humans have only one κ immunoglobulin light chain constant region gene. The encoded proteins differ from one another at several points in the sequence allowing one to determine the origin of a κ light chain based on this information alone. As discussed in Chapter 9, our immune systems are sensitive to these variations and normally generate responses against exogenous antigens while remaining tolerant to endogenous antigens which may differ only subtly. Under certain (poorly understood) circumstances, antibodies specific for self antigens are produced in large quantities and may cause pathologies ranging from mild to severe (fatal). These syndromes are called *autoimmune* (immunity against the self) diseases, and self antigens are called *autoantigens.* Anti-self antigen antibodies are called *autoantibodies*, and B cells producing autoantibodies, as well as T cells participating in autoimmune responses, are often called *autoreactive clones.*

Autoreactive B cell and T cell clones are normal constituents of our immune repertoires throughout our lives. Thus, the key questions in autoimmunity are: what are the mechanisms by which tolerance to self is established and maintained, and how is it lost? In Chapter 9 we examined some of the current concepts regarding establishment and maintenance of self tolerance. Here we shall examine some hypotheses concerning its loss.

Activation of autoreactive clones

Some experiments indicate that autoimmune responses are antigen-driven, as are responses to exogenous antigens. Most autoantigens are proteins or glycoproteins (TD antigens) and must be presented in association with class II MHC molecules in order to elicit a response. Since the vast majority of somatic cells do not express class II antigens, the presentation of many autoantigens remains difficult to understand. Several hypotheses have been advanced to explain the activation of autoreactive clones. These hypotheses are not mutually exclusive and may apply in different situations.

Intracellular antigens

Cytoplasmic antigens are normally unavailable to interact with lymphocyte receptors. Many viruses replicate by budding from cell membranes after having inserted there one or more virus-encoded glycoproteins. In this process, several host cytoplasmic proteins may be brought to the cell surface. In addition, many viral infections induce interferon synthesis which, in turn, can induce the expression of class II MHC antigens on somatic cells. Thus, both requirements of antigen accessibility and presentation may be fulfilled for some intracellular antigens in certain viral infections.

Cross-reactive antigens

Cross-reactivity of antigens borne by infectious agents and self antigens has been documented in several instances. For example, antibodies against the M protein in the cell walls of certain strains of β-hemolytic group A streptococci cross-react with antigens in the myocardium and in the glomerular basement membrane. This cross-reactivity may form the basis of *poststreptococcal glomerulonephritis* and *rheumatic heart disease.*

Altered self antigens

A number of immune hemolytic anemias fall into this category. As described above, some drugs may complex with erythrocyte membrane antigens and stimulate immune responses leading to erythrocyte destruction. Mycoplasma infection of erythrocytes may also lead to production of red cell-reactive antibodies.

Mitogens

Some viruses, such as Epstein-Barr virus (EBV), are mitogenic for human B cells. During EBV infection, it is possible that this mitogenic effect might activate an autoreactive B cell clone. The endotoxins of many gram negative bacteria are also mitogenic and may activate autoantibody production.

General genetic factors

In some instances, a single genetic lesion may lead to autoimmune disease. The tightskin mouse strain which develops cutaneous hyperplasia similar to human scleroderma arose after a mutation on chromosome 2 in the B10.D2 mouse strain. Similarly, the lupus- and rhematoid arthritis-like symptoms in the MRL/*lpr* mouse strain are due to the *lpr* gene (*lpr* = lymphoproliferative). Transfer of this gene into the genetic background of other strains results in the same autoimmune symptomatology.

Strong familial patterns are observed in humans with autoimmune thyroiditis or systemic lupus erythematosus. However, several related individuals may have circulating autoantibodies of the same specificity, yet only one or a few of them may develop overt symptoms. This discordance even extends to monozygotic twins, clearly indicating that genetic factors are only one component in the etiology of autoimmune disease.

Immunogenetic factors

Expression of particular immunoglobulin or T cell receptor genes may be associated with autoimmune disease. The vast majority of human rheumatoid factors are encoded by genes of the V_HI and V_κIIIb gene families. Murine experimental autoimmune encephalitis is caused by T cells specific for myelin basic protein. A majority of these cells express genes of the $V_\beta 8$ family.

Although there may be some preferential use of particular V genes in autoantibodies or T cell receptors, we must consider that foreign and self specificities are derived from the same pool of V genes. Not only are the V genes found in autoreactive clones used in clones specific for foreign antigens, but autoantibodies are even found circulating in individuals with no disease. What then determines the pathogenicity of autoantibodies? It has been suggested that this may be a property of particular D region genes or combinations in association with particular V_H genes.

Many autoantibody V genes show little, if any, somatic mutation compared with known germline gene sequences. Some have suggested that self specificities are germline encoded (though not all germline genes must necessarily encode self specificities). On the other hand, some autoantibodies do appear to show appreciable somatic mutation. Some have even proposed that somatic mutation may be a mechanism by which autoantibodies arise. A single point mutation causing an amino acid change in the V_H may alter specificity of an antibody from phosphorylcholine (a component of bacterial cell walls) to specificity for DNA. However, this does not appear to be a general phenomenon. Overall, the presence or absence of somatic mutation in V genes does not correlate with specificity for self or foreign antigens.

Several diseases with an established or hypothesized autoimmune component are strongly associated with particular HLA haplotypes (see Table 8.III), indicating that they may be under Ir gene control. It is important to keep in mind however, that in addition to reflecting Ir gene control, HLA associations may result from linkage disequilibrium (see Chapter 8).

Altered regulation

Immunodeficiency states are frequently associated with autoimmune processes. Leukemics treated with bone marrow transplantation may produce autoantibodies with various specificities. It is, perhaps, not surprising that such a drastic perturbation of immune system physiology would result in aberrations such as autoimmunity.

Studies of T cell subsets in patients with lupus or primary biliary cirrhosis have shown abnormally high ratios of $CD4^+$ ($T_{h/i}$)/$CD8^+$ ($T_{c/s}$) cells. This apparent decrease in $T_{c/s}$ cells suggests that active suppression may be an important mechanism in the maintenance of tolerance to self, and that loss of suppression leads to autoimmune disease.

Idiotype-anti-idiotype interactions may also play a role in initiating autoimmune phenomena. Patients with myasthenia gravis produce antibodies to skeletal muscle acetylcholine receptors (AchR), and also produce anti-idiotypes against these antibodies. Pregnant women with the disease may pass these antibodies to their fetuses and cause a transient or long-lasting myasthenia syndrome via anti-idiotype activation of fetal anti-AchR clones. Production of anti-thyroglobulin antibodies in mice may also be stimulated with anti-idiotypic antibodies.

Idiotype network interactions might activate autoreactive clones by (at least) two mechanisms. Antibodies produced in a response against a foreign antigen may cross-react with idiotopes on antibodies with self-specificity and activate B cells. Alternatively, anti-idiotypes produced in an immune response may activate self-specific B cells sharing idiotopes with antibodies against the foreign antigen. Idiotype-bearing or anti-idiotypic T cells may also play a role in these mechanisms.

Autoimmune diseases mediated by antibodies

The harmful effects of autoantibodies were first demonstrated when it was shown that the sera of patients with hemolytic anemia contained hemolysing antibodies. Ehrlich used the phrase *horror autotoxicus* when he postulated that autoantibodies were intrinsically pathogenic. This is clearly not correct since we now know that all of us produce some autoantibodies at various times in our lives. When proposing an autoimmune etiology for a particular pathology, several key points should be clearly demonstrated:

1. Presumed autoantigens should be clearly identified and characterized.

2. The specificity of the autoreactive antibodies and lymphocytes should be evident in appropriate controlled *in vitro* experiments.

3. Autoantigen-specific antibodies and/or lymphocytes should be identified within lesions.

4. Autoantibodies or autoreactive cells should produce similar symptoms when transferred into other animals, or they should produce characteristic histologic alterations in tissue or organ culture systems.

5. Animals immunized with the autoantigen should exhibit pathology characteristic of the disease.

Autoantibodies may be classified into three major groups. As the name implies, *organ-specific autoantibodies* bind antigens associated with one organ system, and are responsible for the symptomatology affecting that system. Examples include anti-acetylcholine receptor antibodies in myasthenia gravis, and anti-desmosome antibodies in pemphigus.

Non-organ-specific autoantibodies are usually found in systemic autoimmune diseases. Examples include antibodies against single or double-stranded DNA, or ribonucleoprotein antigens. These autoantibody specificities are observed in systemic lupus erythematosus. Antimitochondrial autoantibodies are not organ-specific, yet they are found in a large fraction of patients with primary biliary cirrhosis.

Multispecific autoantibodies are capable of binding to more than one self antigen. It has been suggested that under normal circumstances, these autoantibodies might prevent autoimmune disease through a "housekeeping" role, complexing with defunct proteins and senescent cells and promoting their ingestion and destruction by cells of the mononuclear phagocyte system.

Autoantibodies may cause pathology in different ways. As we have described above, antibody binding to cell surfaces or tissue membrane or matrix components may activate ADCC and complement. Autoantibodies may bind to cell-surface receptors and block the physiological effect of the ligand (hormone, factor, etc.), or they may act as receptor agonists. The latter may mimic overproduction of the particular ligand. These situations are the "stimulatory reactions" (type V) in our discussion.

Autoantibody-mediated tissue destruction

Autoimmune hemolytic anemias. These clinically heterogeneous syndromes are grouped together under the name *acquired hemolytic disease* (*AHD*).

They have in common the production of autoantibodies to red cell antigens and hemolysis (type II reactions). Common symptoms include anemia, jaundice (hyper bilirubinemia due to increased degradation of hemoglobin), and splenomegaly due to uptake of erythrocyte ghosts by splenic macrophages.

These diseases are classified according to the properties of the hemolytic autoantibodies. In some conditions, the antibodies bind red cell membranes at 37°C, these are called *warm reactive antibodies*. Diseases caused by this class of hemagglutinin are classified as either idiopathic, or secondary AHD. Warm reactive hemagglutinins are frequently associated with lymphoproliferative diseases (such as chronic lymphocytic leukemia), systemic lupus erythematosus, and some malignancies and infections. In these situations the AHD is secondary. If no other disease state is identified, the AHD is considered idiopathic.

Some hemagglutinins react poorly with red cells at 37°C, but well at temperatures below 32°C. These are called *cold reactive antibodies* or *cold agglutinins*. AHD caused by these antibodies is called *paroxysmal cold hemoglobinuria* or *cold agglutinin disease*. In both of these conditions, hemolysis only occurs when the body, or a part of it is exposed to cold. Cold agglutinins have been shown to use the same family of light chain V genes (V_{κ}IIIb) as rheumatoid factors.

Other cytopenias may also be the result of autoantibody production. *Idiopathic thrombocytopenic purpura (ITP)* results from platelet desctruction due to anti-platelet autoantibodies. Decreased blood clotting efficiency is manifested by mucocutaneous purpural lesions. *Autoimmune neutropenia* is caused by antibodies reactive with neutrophil surface antigens. Myeloid hyperplasia indicates compensation for decreased neutrophil life span in the circulation.

Autoimmune thyroiditis, also known as *Hashimoto's disease*, is characterized by thyroid enlargment (goiter) and varying degrees of functional impairment. Antibodies against thyroglobulin, thyroid peroxidase (a microsome constituent), and thyroid hormones have been detected in affected individuals. Histologically, a marked lymphocytic infiltration of the thyroid is seen. Some regions resemble a DTH reaction (prominent mononuclear cell component) suggesting that cell-mediated immunity also contributes to the etiology of this disease.

The diseases grouped under the name *pemphigus* are all characterized by antibodies binding the protein *desmoglin*, a component of the desmosomes connecting skin keratinocytes. The most prominent symptom is the eruption of fragile blisters (bullae) of varying sizes. The pathogenicity of the anti-desmoglin antibodies was demonstrated by injection of serum IgG from pemphigus patients into newborn mice, which developed the characteristic skin lesions.

Idiopathic adrenal atrophy is a form of hypoadrenalism (the general name for which is *Addison's disease*) which may be caused by an autoimmune mechanism. Antibodies reacting with microsomes in cells of all three layers of the adrenal cortex are detected in many patients. This disease may be induced experimentally by immunizing animals with autologous adrenal tissue homogenized in Freund's adjuvant. Despite this evidence, some debate exists whether these autoantibodies initiate pathology, or if they arise subsequently to adrenal cell destruction by other means, with the liberation of previously sequestered antigens.

Several other pathological processes may have an organ-specific autoantibody component of their etiologies. The syndrome known as *antitubular basement membrane (anti-TBM) antibody disease* (also called *tubulointerstitial nephritis*) is an inflammation of the proximal tubular interstitium whose origin is described in its name. *Vitiligo* is the depigmentation of fairly large (several cm^2 in area) irregular patches of skin. An autoimmune etiology is strongly suspected since antibodies against melanocytes have been detected in patients. In *primary ovarian failure* amenorrhea due to cessation of ovarian steroid synthesis may be caused by antibodies reactive to the theca interna of ovarian follicles.

Autoantibody-mediated receptor blockade

The cardinal symptom of *myasthenia gravis* is generalized muscle weakness, the result of autoantibody binding to acetylcholine receptors on skeletal muscle. Not only does antibody binding prevent the normal action of the agonist, acetylcholine, but it also causes removal of receptors from the cell membrane. Serum antibodies from myasthenics reproduce the symptoms of muscle weakness when injected into rats.

The action of acetylcholine (Ach) is normally terminated by acetylcholinesterase (AchE) at the neuromuscular junction. Thus, AchE inhibitors increase the concentration of Ach at the synapse. In normal individuals, this results in persistent myocyte membrane depolarization, and overall muscle weakness. In myasthenics, however, the increased amounts of Ach compete with antibodies for the receptor, and AchE inhibitors *increase* muscle strength. This phenomenon is one of the simplest clinical tests for myasthenia, and is the basis of therapy for this disease. Tensilon is an AchE inhibitor, and a test for myasthenia gravis is often referred to as a *Tensilon test*.

Pernicious anemia consists of gastric mucosal dysfunction and atrophy combined with megaloblastic anemia and neurologic manifestations. These symptoms result from production of antibodies binding gastric parietal cells and intrinsic factor. Intrinsic factor is a secretory product of neck cells in the gastric glands, and is required for absorption of vitamin B_{12}, a cofactor essential to a number of biochemical pathways. Antibodies binding intrinsic factor inhibit B_{12} uptake. Ensuing biochemical derangements together with gastric atrophy are responsible for the spectrum of symptoms associated with this disease.

A few other conditions have been associated with autoantibody-receptor blockage. *Type B insulin resistance* is due to circulating autoantibodies to insulin and/or the insulin receptor. Many of these patients also have warty growths and hyperpigmentation of areas of skin (acanthosis nigricans). These patients also often have autoantibodies with other specificities such as nuclear antigens and DNA.

Autoimmune mechanisms may also be responsible for disease in some patients with *asthma.* Autoantibodies binding bronchial β-adrenergic receptors have been found in some asthmatics. These may inhibit the bronchodilating action of β-adrenergic agonists.

Autoantibody-mediated receptor stimulation

Grave's disease is closely related to Hashimoto's disease. Symptomatically, Grave's disease may (sometimes) be distinguished from Hashimoto's disease by the tendency in the former toward hyperthyroidism and proptosis. Microscopic examination of the thyroid shows follicular cell hyperplasia, in addition to lymphocytic infiltration. Both antibody and cell-mediated autoimmune reactions have been implicated in Grave's disease. Antibodies with the specificities found in Hashimoto's thyroiditis are present, with the important addition of antibodies binding receptors for thyroid stimulating hormone (TSH). These antibodies have been called *long-acting thyroid stimulator* (*LATS*). Binding of LATS to the TSH receptor causes increases in intracellular cyclic AMP and may induce follicular cell proliferation. Cellular immunity also plays a role in tissue destruction in Grave's disease.

Cushing's syndrome results from any process leading to overproduction of adrenal corticosteroids. Although most cases are due to hormone-secreting neoplasms, some have been ascribed to production of auto-antibodies binding receptors for adrenocorticotrophic hormone (ACTH) and stimulating cortical cells.

Autoimmune diseases mediated by T cells

Experimental allergic encephalitis (*EAE*) is a model of a principally T cell-mediated tissue destructive process. The disease is induced in mice by immunization with brain homogenate or myelin basic protein in Freund's adjuvant. Inflammatory foci and demyelination in the white matter appear approximately three weeks after injection. Transfer of T cells will reproduce the lesions in naive (untreated) mice, while transfer of serum or purified antibodies has no effect. *Acute disseminated encephalitis* is very similar to EAE. This condition may occur after rabies vaccination. The vaccine is prepared from homogenates of infected brain tissue. Brain lesions consist of perivascular lymphocytic infiltrations in the white matter, much as in EAE. Autoantibodies to brain antigens are also produced.

Type I diabetes is also referred to as insulin-dependent diabetes mellitus (IDDM). One of the suspected causes of this disease is an autoimmune reaction against the beta (insulin-secreting) cells of the pancreatic islets of Langerhans. Insulitis is observed microscopically as a mononuclear infiltrate. At presentation, most patients with IDDM have circulating antibodies reactive with islet cells. However, a cell-mediated reaction appears to be the predominant mechanism of islet cell destruction. Class II HLA antigens have been detected on beta cells in IDDM, and $CD8^+$ T cells are prominent in leukocytic infiltrates of the islets. T cells from the BB/W rat strain, prone to IDDM, cause IDDM when transferred into normal rats. Furthermore, neonatal thymectomy can delay the onset of disease or prevent it entirely. Treatment with cyclosporine, a powerful immunosuppressant, is occasionally successful in ameliorating IDDM.

Progressive systemic sclerosis, also called *scleroderma*, is a progressive fibrosis of connective tissue throughout the body caused by overproduction of several types of collagen. The process usually begins in the skin, but eventually affects other tissues. Death is often due to loss of renal, cardiac, or pulmonary function. A lymphocytic infiltrate often precedes the fibrotic changes in affected tissue. T cells in these infiltrates produce IL-1 which stimulates production of collagen by fibroblasts. Autoantibodies are also produced in scleroderma. Specificities include topoisomerase I, centromere, RNA polymerase and, occasionally, DNA.

Autoantibodies against retinal cells have been described in patients with several different forms of *uveitis*. However, T cells are hypothesized to play the more prominent role in tissue destruction. The predominant antigen specificity is the interphotoreceptor retinoid binding protein.

Systemic autoimmune diseases

This group of syndromes, with widely varying symptomatologies, is associated with high levels of circulating autoantibodies specific for antigens distributed throughout the body. Not all tissues containing these antigens are affected, and it is not well-established that these autoantibodies initiate the disease, or if they are elicited after liberation of antigens following some other destructive process. Circulating autoantibodies form immune complexes in the blood, which often deposit in the kidney causing glomerulonephritis. In contrast to organ-specific autoantibodies, only rarely can one initiate pathology by transfer of non-organ-specific autoantibodies into a healthy individual. While there are specific clinical features which have "classically" been associated with particular systemic autoimmune diseases, there is a great degree of "overlap," many patients having signs and symptoms and autoantibody specificities of more than one syndrome.

Systemic lupus erythematosus (*SLE*) is a syndrome which most often afflicts women (9:1 over men), and may present in a variety of ways, and follow numerous courses. Any organ system may be affected, however, most often involved are the skin (patchy erythema), joints (arthritis), and kidneys (glomerulonephritis). Some of the antibody specificities associated with SLE are: single or double-stranded DNA, ribonucleoproteins (called Ro or SSA, La or SSB, and Sm), and histones. Antibodies which bind nuclear antigens are called *anti-nuclear antibodies* (*ANA*). Polyspecific autoantibodies capable of binding to several autoantigens have also been found. Approximately 15% of lupus patients have autoantibodies binding ribosomal phosphoproteins. These autoantibodies have been shown to inhibit protein translation *in vitro*. Autoantibodies are found in immune complexes in the skin and kidney, however, autoantibodies from animal models of lupus, such as (NZB X NZW)F_1 mice, fail to cause pathology when injected into normal mice. Autoantibody production, of some specificities at least, appears to require T cell help. Human $TCR2^+$ (α/β), $CD4^-/CD8^-$ (double-negative) T cells (along with $CD4^+$) cells have been shown to promote synthesis of IgG anti-DNA autoantibodies. Whether this double-negative T cell population plays a role in pathogenesis of SLE remains to be seen. There are no particular V genes which characterize autoantibody or autoreactive T cell specificities in SLE. Both germline encoded and somatically mutated autoantibodies have been found. The events leading to activation of autoreactive clones in SLE remain unknown.

Rheumatoid arthritis (*RA*) affects primarily the synovia and cartilage of various joints, but may involve the skin, musculoskeletal, circulatory and respiratory systems. Inflammatory lesions occurring in tissues other than synovia are called *rheumatoid nodules*. With very few exceptions, the sera of RA patients contain large amounts of *rheumatoid factors* (*RF*), antibodies specific for the F_c fragment (Gm allotypic determinants) of IgG. Additional autoantibody specificities identified in some patients include nuclear antigens and collagen type II (CII). CII-specific autoreactive clones appear to be important in RA. CII-specific autoantibodies and T cells persist in the joints of RA patients over many years. The CII-specific T cells secrete TNFs α and β which induce collagen degradation. IL-1 and IL-6 are found in high amounts in the synovial fluid of RA patients. Furthermore, the injection of CII with adjuvants into animals can cause a synovitis/arthritis resembling RA. This pathology can be transferred to healthy animals by antibodies or T cells. "Arthritogenic" antibodies are predominantly of the IgG3 isotype. This class of antibodies binds complement very effectively. RA has been associated with various HLA class II specificities (see Chapter 8).

Sjögren's syndrome is a chronic autoimmune disease affecting principally the salivary and lacrimal glands, and is associated with rheumatoid arthritis in about 50% of cases. The loss of secretions from the affected glands results in xerophthalmia (dry eyes) and xerostomia (dry mouth). Sjögren's syndrome is also associated frequently with SLE and other immune disorders. Commonly detected antibody specificities include salivary gland antigens, ribonucleoproteins (as in SLE), and rheumatoid factors. The highest levels of ribonucleoprotein-specific autoantibodies have been associated with HLA DQw1 and DQw2. A role for cell-mediated immunity has also been implicated in the etiology of this syndrome.

Polymyositis and dermatomyositis are similar entities in the spectrum of systemic autoimmune diseases. Skeletal muscle weakness and pain are the cardinal symptoms of polymyositis (PM). Cardiac muscle and smooth muscle (especially of the gastrointestinal tract) may also be affected. When cutaneous lesions are the predominant symptom, the diagnosis of dermatomyositis (DM) is made. The most common findings are a heliotrope rash, Gottron's sign (exfoliative papular lesions on joint extensor surfaces), and a macular rash on areas of skin exposed to sunlight. Lungs, joints, and blood vessels may also be affected in these syndromes. An increased incidence of malignancy is associated with these diseases, especially DM. Autoantibodies binding myoglobin and myosin are present in most affected individuals. About 25% of patients also have autoantibodies binding either histidyl-, threonyl-, or alanyl-tRNA synthetases. Antibodies binding a poorly characterized antigen known as Mi-2 characterize about 10% of patients with DM and without PM. Anti-mitochondrial and cytoskeletal autoantibodies have also been found in a large fraction of patients. Cell-mediated immunity has also been implicated in the muscle and skin damage in these diseases.

In *Goodpasture's syndrome*, autoantibodies are formed against antigens found in pulmonary and glomerular basement membranes, possibly collagen type IV. Interestingly, antibodies eluted from glomerular basement membranes fail to bind to lung basement membranes and *vice versa*. Complement activation and cellular cytotoxicity lead to pneumonitis and glomerulonephritis.

Primary biliary cirrhosis is a chronic granulomatous inflammatory process destroying the intrahepatic biliary system. Other organs which may be affected include the pancreas, and salivary and lacrimal glands. These patients have a marked reduction in the number of circulating $CD8^+$ cells (cytotoxic/suppressor phenotype), and a diminished ability to switch from IgM to IgG production. The characteristic finding is circulating antibodies against mitochondrial antigens, specifically, the acyltransferases of the pyruvate dehydrogenase and α-ketoacid dehydrogenase complexes.

Since the events initiating spontaneous autoimmune processes are not known, we are unable to take action to prevent or reverse them. In a few instances, specific therapies are effective. For example, acetylcholinesterase inhibitors for myasthenics, or thyroid hormone supplements for patients with Hashimoto's disease. In systemic autoimmune diseases such as systemic lupus erythematosus, immunosuppressive therapy is often the only means available for prolonging life. Corticosteroids and antineoplastic (cytotoxic) agents are used in this manner. The greatest danger of immunosuppressive therapy is an increased susceptibility to infection.

ABNORMAL LEUKOCYTE PROLIFERATION

Uncontrolled leukocyte proliferation may be due either to the derangement of normal regulatory mechanisms limiting cellular division, or a neoplastic transformation event. These diseases may be broadly

Table 10.II. DISEASES OF ABNORMAL LEUKOCYTE PROLIFERATION

Acute lymphocytic leukemia
Chronic lymphocytic leukemia
Acute myelogenous leukemia
Chronic myelogenous leukemia
Hodgkin's disease
Non-Hodgkin's lymphoma
Multiple myeloma
Waldenström's macroglobulinemia
Essential thrombocythemia
Myeloid metaplasia

categorized into the lymphomas and leukemias. *Lymphomas* are solid masses (tumors) consisting of lymphoid or myeloid cells. When proliferating cells enter the circulation, and label *leukemia* is appropriate. These diseases may also be classified with respect to the phenotype of the proliferating cells, stem cells, B cells, T cells, monocytes, etc. Table 10.II lists the major categories of leukocyte proliferative diseases. Some of these have been described in other chapters. We will not present a detailed discussion of each of these disorders, a few points of interest are presented below.

Hodgkin's disease (also known as *Hodgkin's lymphoma*), is a group of disease entities which may have different etiologies, but have in common a single feature: the appearance of an abnormal cell type called the *Reed-Sternberg cell.* These cells are large with an amphophilic cytoplasm, and have a characteristic bilobed or double nucleus with prominent nucleoli. The origin of this cell is not known, different investigators have argued its genesis from B cells, T cells, or macrophages. Four histologic types of Hodgkin's disease are distinguished: *lymphocyte predominance*, *lymphocyte depletion*, *mixed cellularity* and *nodular sclerosis.* These differ in the cellular composition of tumors, degree of malignancy, and in average age of onset. Some researchers hypothesize that Hodgkin's disease begins as an inflammatory reaction to an (as yet) unidentified infectious agent. Treatment is most often by a combination of radio- and chemotherapy.

The molecular basis of *Burkitt's lymphoma* has been extensively investigated. Early cytogenetic studies showed that in a high percentage of affected individuals, the malignant cells had a translocation involving chromosomes 8 and 14 [t(8,14)]. Subsequent research has shown that this translocation occurs in the area of chromosome 14 containing the immunoglobulin heavy chain locus, and the region of chromosome 8 where the *c-myc* oncogene resides. The transcriptionally active heavy chain locus induces transcription of the oncogene. The events leading to the translocation, the function of the *myc* protein and the mechanism of oncogenesis remain to be elucidated. Translocations involving κ and λ light chain gene loci with *c-myc* have also been described. The highest incidence of Burkitt's lymphoma is in Africa where a very strong association with Epstein-Barr virus (EBV) infection has been documented. Histologically and clinically identical forms of the disease occurring in other parts of the world appear not to be associated with EBV. Burkitt's lymphoma usually presents as a rapidly growing mass in the mandible or other facial bone, or in the abdomen. Progress of the tumors may be rapid even with treatment. Five year survival is approximately 30%.

Chronic myelogenous leukemia (*CML*) is generally slow in onset and slow to progress. Leukemic infiltration of bone marrow leads to anemia, while massive hepatosplenomegaly results from accumulation of leukemic cells in the liver and spleen. As with Burkitt's lymphoma, this neoplasm is associated with a characteristic chromosomal aberration. Ninety percent of CMLs have a translocation involving chromosomes 9 and 22 [t(9,22)]. This has become known as the *Philadelphia chromosome*. This translocation involves the *c-abl* oncogene, the product of this gene is a tyrosine kinase. The mechanism of oncogenesis has not been determined.

Recent studies of *chronic lymphocytic leukemia* (*CLL*) indicate that as many as 80% of CLLs of B cell origin express the CD5 antigen. This cytodifferentiation antigen has also been associated with autoreactive B cell clones.

IMMUNODEFICIENCY

Listed in Table 10.III are the various ways in which a state of immunodeficiency may arise. The *immunocompromised* individual is principally concerned with infectious microorganisms in the environment. In addition to those organisms which may cause disease in healthy people, immunodeficient hosts must also be concerned with usually non-pathogenic organisms which live as commensals in and on our bodies, or are encountered with high frequency in the environment. These microbes normally are not virulent and do not proliferate sufficiently to cause disease in the context of a healthy immune system. However, they may multiply unchecked in the immunodeficient host. For this reason, these are referred to as *opportunistic infections* (see Table 7.III). Infectious pneumonia is a major cause of death in

Table 10.III. CAUSES OF IMMUNODEFICIENCY

Genetic defect in lymphopoiesis or function of mature B and/or T cells, e.g., SCID (see Chapter 2)
Genetic defect in development of lymphoid organs, e.g., congenital thymic aplasia (see Chapters 2 and 7)
Genetic defect in expression of MHC antigens, e.g., bare lymphocyte syndrome (see Chapter 8)
Infectious diseases, e.g., measles, tuberculosis, leprosy, AIDS (see Chapter 7)
Chronic systemic disease, e.g., diabetes mellitus, chronic renal insufficiency
Malignancy, e.g., Hodgkin's disease, lymphoma, leukemia, end-stage solid tumors
Immune system dysregulation, e.g., common variable immunodeficiency (see Chapter 6), autoimmune diseases, old age?
Common drug side effects, e,g., cytotoxic cancer chemotherapeutic agents
Idiosyncratic drug side effects
Immunosuppressive therapy in autoimmune disease or transplantation
Malnutrition

immunodeficiency states. Even massive amounts of antimicrobial agents are often unsuccessful in combating infection in the absence of immune system function.

SOURCES AND SUGGESTED ADDITIONAL READING

Acha-Orbea, H., Mitchell, D. J., Timmermann, L., Wraith, D. C., Tausch, G. S., Waldor, M. K., Zamvil, S. S., McDevitt, H. O. & Steinman, L. (1988) Limited heterogeneity of T cell receptors from lymphocytes mediating autoimmune encephalomyelitis allows specific immune intervention. *Cell*, **54**:263–273.

Andrada, J. A., Murray, F. T., Andrada, E. C. & Ezrin, C. (1979) Cushing's syndrome and autoimmunity. *Arch. Pathol. Lab. Med.*, **103**:244–246.

Arnett, F. C., Goldstein, R., Duvic, M. & Reveille, J. D. (1988) Major histocompatibility complex genes in systemic lupus erythematosus, Sjögren's syndrome, and polymyositis. *Am. J. Med.*, **85** (Suppl. 6A):38–41.

Betterle, C., Zanette, F., Zanchetta, R., Pedini, B., Trevisan, A., Mantero, F. & Rigon, F. (1983) Complement-fixing autoantibodies as a marker for predicting onset of idiopathic Addison's disease. *Lancet*, **i**:1238–1241.

Bona, C. A. (1988) V genes encoding autoantibodies: molecular and phenotypic characteristics. *Annu. Rev. Immunol.*, **6**:327–358.

Bystryn, J. C. & Pfeffer, S. (1988) Vitiligo and antibodies to melanocytes. *Prog. Clin. Biol. Res.*, **256**:195–206.

Cairns, E., Kwong, P. C., Misener, V., Ip, P., Bell, D. A. & Siminovich, K. A. (1989)

Analysis of variable region genes encoding a human anti-DNA antibody of normal origin. *J. Immunol.*, **143**:685–691.

Camussi, G., Kitazawa, K. & Andres, G. (1985) Mechanisms of tissue injury induced by immune complexes formed in the circulation or "*in situ*". *Diagn. Immunol.*, **3**:109–118.

Conn, D. A. (1986) Intrinsic factor antibodies in relation to disease. *Med. Lab. Sci.*, **43**:220–224.

Dameshek, W. & Schwartz, S. O. (1938) The presence of hemolysins in acute hemolytic anemia. *New Engl. J. Med.*, **218**:75–80.

Davidson, A., Shefner, R., Livneh, A. & Diamond, B. (1987) The role of somatic mutation of immunoglobulin genes in autoimmunity. *Annu. Rev. Immunol.*, **5**:85–108.

Dotta, F. & Eisenbarth, G. S. (1989) Type I diabetes mellitus: a predictable autoimmune disease with interindividual variation in the rate of beta cell destruction. *Clin. Immunol. Immunopathol.*, **50**:S85–95.

Foon, K. A. & Gale, R. P. (1987) Immunologic classification of lymphoma and lymphoid leukemia. *Blood Rev.*, **1**:77–88.

Frasser, C. M. & Venter, J. C. (1982) Autoantibodies to beta 2-adrenergic receptors and allergic respiratory disease. *Surv. Immunol. Res.*, **1**:365–370.

Fregeau, D. R., Davis, P. A., Danner, D. J., Ansari, A., Coppel, R. L., Dickson, E. R. &. Gershwin, M. E. (1989) Antimitochondrial antibodies of primary biliary cirrhosis recognize dihydrolipoamide acyltransferase and inhibit enzyme function of the branched chain α-ketoacid dehydrogenase complex. *J. Immunol.*, **142**:3815–3820.

Fujinami, R. S. (1988) Virus-induced autoimmunity through molecular mimicry. *Ann. N. Y. Acad. Sci.*, **540**:210–217.

Goñi, F. R., Chen, P. P., McGinnis, D., Arjonilla, M. L., Fernandez, J., Carson, D., Solomon, A., Mendez, E. & Frangione, B. (1989) Structural and idiotypic characterization of the L chains of human IgM autoantibodies with different specificities. *J. Immunol.*, **142**:3158–3163.

Hamburger, R. N. (1975) Peptide inhibition of the Prausnitz-Küstner reaction. *Science*, **189**:389–390.

Hemachudha, T., Griffin, D. E., Giffels, J. J., Johnson, R. T., Moser, A. B. & Phanuphak, P. (1987) Myelin basic protein as an encephalitogen in encephalomyelitis and polyneuritis following rabies vaccination. *N. Engl. J. Med.*, **316**:369–374.

Hirose, S., Tanaka, T., Nussenblatt, R. B., Palestine, A. G., Wiggert, B., Redmond, T. M., Chader, G. J. & Gery, I. (1988) Lymphocyte responses to retinal-specific antigens in uveitis patients and healthy subjects. *Curr. Eye. Res.*, **7**:393–402.

Holdsworth, S., Boyce, N., Thomson, N. M. & Atkins, R. C. (1985) The clinical spectrum of acute glomerulonephritis and lung hemorrhage (Goodpasture's syndrome). *Q. J. Med.*, **55**:75–86.

Klippel, J. H., ed. (1988) Systemic lupus erythematosus. *Rheum. Dis. Clin. North Am.*, **14**(1).

Korman, N. (1988) Pemphigus, *J. Am. Acad. Dermatol.*, **18**:1219–1238.

Krieg, T. & Meurer, M. (1988) Systemic scleroderma. Clinical and pathophysiologic aspects. *J. Am. Acad. Dermatol.*, **18**:457–481.

LaBarbera, A. R., Miller, M. M., Ober, C. & Rebar, R. W. (1988) Autoimmune etiology in premature ovarian failure. *Am. J. Reprod. Immunol. Microbiol.*, **16**:115–122.

Lilja, G., Sundin, B., Graff-Lonevig, V., Hedlin, G., Heilborn, H., Norrlind, K., Pegelow, K. O. & Lowenstein, H. (1989) Immunotheraphy with cat- and dog-dander extracts. IV. Effects of 2 years of treatment. *J. Allergy Clin. Immunol.*, **83**:37–44.

Marone, G., Casolaro, V., Cirillo, R., Stellato, C. & Genovese, A. (1989) Pathophysiology of human basophils and mast cells in allergic disorders. *Clin. Immunol. Immunopathol.*, **50**: S24–40.

McCluskey, R. T. (1987) Immunopathogenetic mechanisms in renal disease. *Am. J. Kidney Dis.*, **10**: 172–180.

Murano, G. (1978) The "Hageman" connection: interrelationships of blood coagulation, fibrino(geno)lysis, kinin generation, and complement activation. *Am. J. Hematol.*, **4**: 409–417.

Nakamura, R. M. & Binder, W. L. (1988) Current concepts and diagnostic evaluation of autoimmune disease. *Arch. Pathol. Lab. Med.*, **112**: 869–877.

Natvig, J. B. & Winchester, R. J., eds. (1988) Rheumatoid arthritis. *Springer Semin. Immunopathol.*, **10**(2/3).

Newland, A. C. (1987) Idiopathic thrombocytopenic purpura and IgG: a review. *J. Infect.*, **15** (Suppl. 1): 41–49.

Newsom, D. J. (1988) Autoimmunity in neuromuscular disease. *Ann. N. Y. Acad. Sci.*, **540**: 25–38.

Patten, E. (1987) Immunohematologic diseases. *JAMA*, **258**: 2945–2951.

Ricci, M. (1989) Immunoregulation in clinical diseases: an overview. *Clin. Immunol. Immunopathol.*, **50**: S3–12.

Sanz, I., Hwang, L.-Y., Hasemann, C., Thomas, J., Wasserman, R., Tucker, P. & Capra, J. D. (1988) Polymorphisms of immunologically relevant loci in human disease. Autoimmunity and human heavy chain variable regions. *Ann. N. Y. Acad. Sci.*, **546**: 133–142.

Schwartz, R. C. & Witte, O. N. (1988) The role of multiple oncogenes in hematopoietic neoplasia. *Mutat. Res.*, **195**: 245–253.

Selenkow, H. A., Wyman, P. &. Allweiss, P. (1984) Autoimmune thyroid disease: an integrated concept of Graves' and Hashimoto's diseases. *Compr. Ther.*, **10**: 48–56.

Shivakumar, S., Tsokos, G. C. & Datta, S. K. (1989) T cell receptor α/β expressing double-negative ($CD4^-/CD8^-$) and $CD4^+$ T helper cells in humans augment the production of pathogenic anti-DNA autoantibodies associated with lupus nephritis. *J. Immunol.*, **143**: 103–112.

Sontheimer, R. D., ed. (1988) Polymyositis/dermatomyositis. *Clin. Dermatol.*, **6**(2).

Stacy, D. W., Skelley, S., Watson, T., Elkon, K., Weissbach, H. & Brot, N. (1988) The inhibition of protein synthesis by IgG containing anti-ribosome P autoantibodies from systemic lupus erythematosus patients. *Arch. Biochem. Biophys.*, **267**: 398–403.

The third international conference on rheumatic fever and rheumatic heart disease. (1988) *N. Z. Med. J.*, **101**(846).

Tomer, Y. & Schoenfeld, Y. (1988) The significance of natural autoantibodies. *Immunol. Invest.*, **17**: 389–424.

Tsushima, T., Omori, Y., Murakami, H., Hirata, Y. & Shizume, K. (1989) Demonstration of heterogeneity of autoantibodies to insulin receptors in type B insulin resistance by isoelectric focusing. *Diabetes*, **38**: 1090–1096.

Whittaker, S. J., Dover, J. S. & Greaves, M. W. (1986) Cutaneous polyarteritis nodosa associated with hepatitis B surface antigen. *J. Am. Acad. Dermatol.*, **15**: 1142–1145.

Youinou, P. & Pennec, Y. (1987) Immunopathologic features of primary Sjögren's syndrome. *Clin. Exp. Rheumatol.*, **5**: 173–184.

Epilogue

We have seen that the immune system is a double-edged sword. While it protects us from infection, it may also disrupt the functions of other body systems while performing its defensive role. At times the immune system may even become so deranged as to attack our own tissues. Hopefully, as we understand more clearly the principles of immune system function and regulation we will be able to prevent or reverse these harmful processes.

Research in immunology has yielded many tremendously important technologies in medicine. Vaccination is one of the most important contributions of modern medical science to human health and disease prevention worldwide. Life-saving practices such as blood transfusion and organ transplantation would be impossible without knowledge of immune system function and the means to suppress it. Monoclonal antibodies have become critically important tools in almost every biological science and have revolutionized the clinical diagnosis and monitoring of a multitude of diseases.

Utilization of modern molecular biological techniques by immunologists has opened the door to a deeper understanding of the genetic mechanisms governing immune system function and dysfunction. Without a doubt, we have merely scratched the surface of immunology, which will remain an exciting and truly interdisciplinary area of medical research for many years to come. We hope to have presented an introduction which will make the reader conversant with the modern concepts of immunology, and stimulate at least a few to pursue further study and make their own contributions to this field.

Index

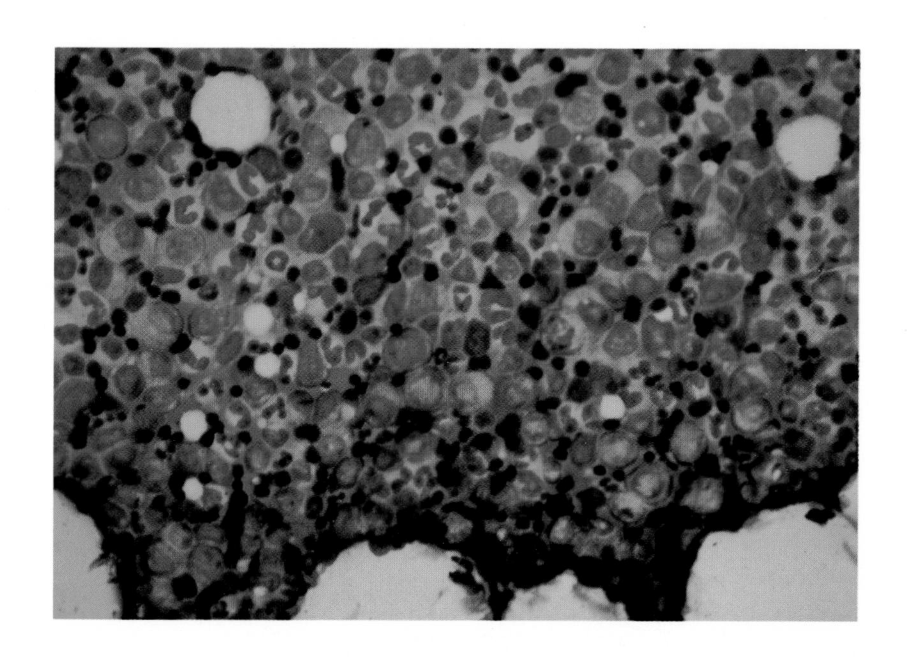

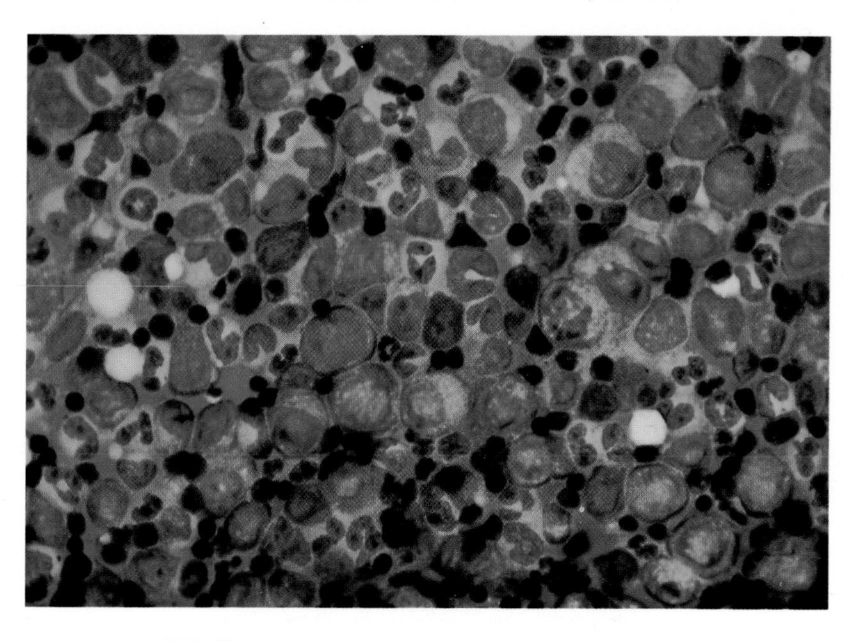

COLOR PLATE I. See Chapter 2, page 7, Figure 2.2.

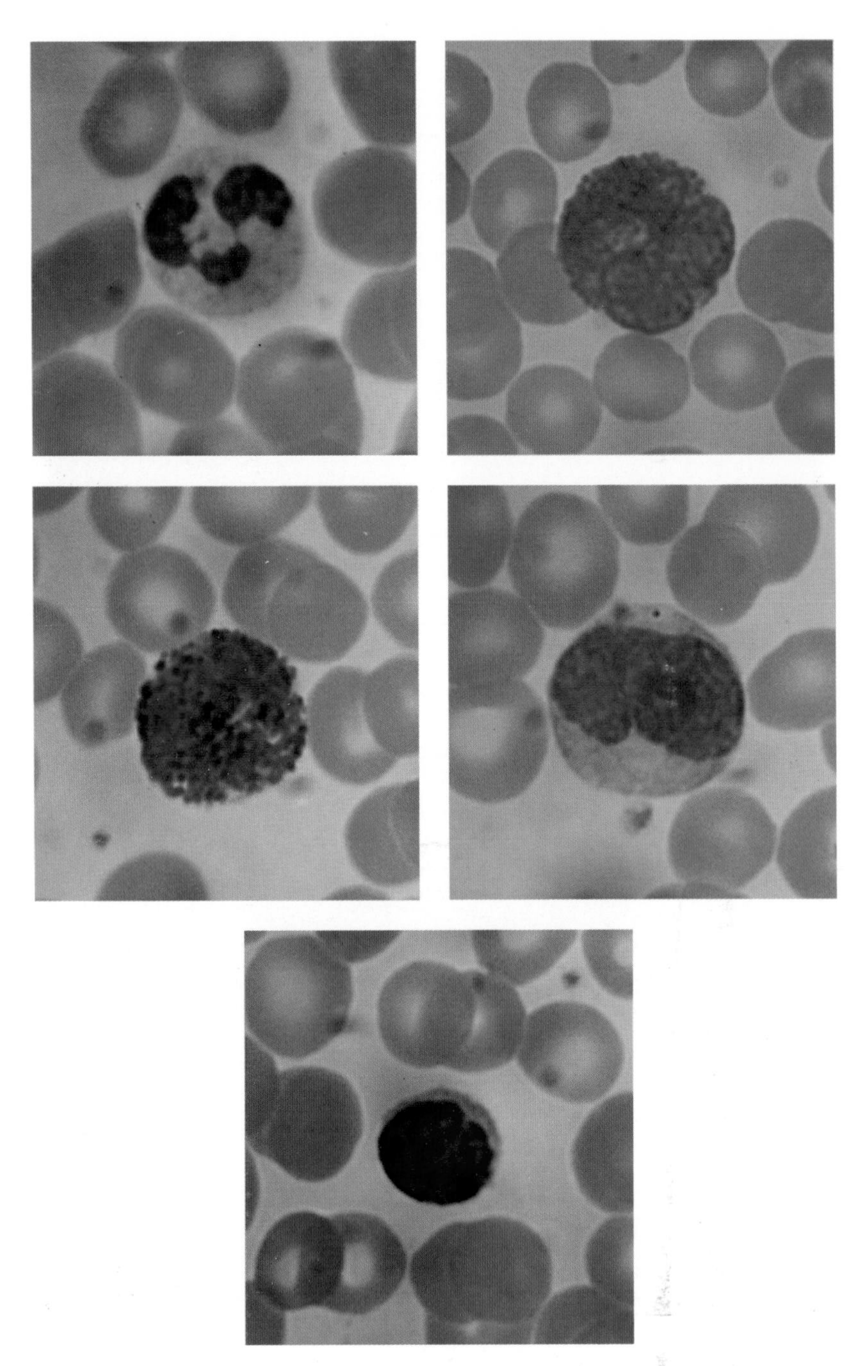

COLOR PLATE II. See Chapter 2, page 14, Figure 2.4.

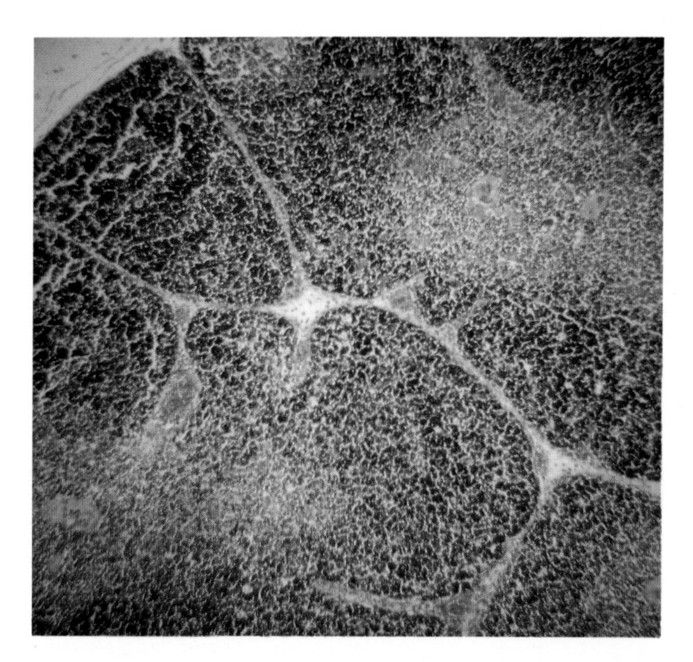

COLOR PLATE III. See Chapter 2, page 24, Figure 2.11a.

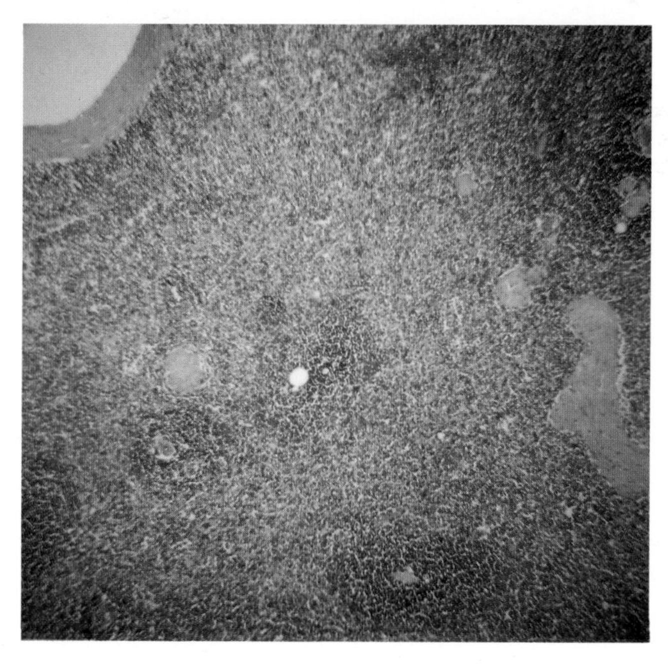

COLOR PLATE IV. See Chapter 2, page 30, Figure 2.14a.

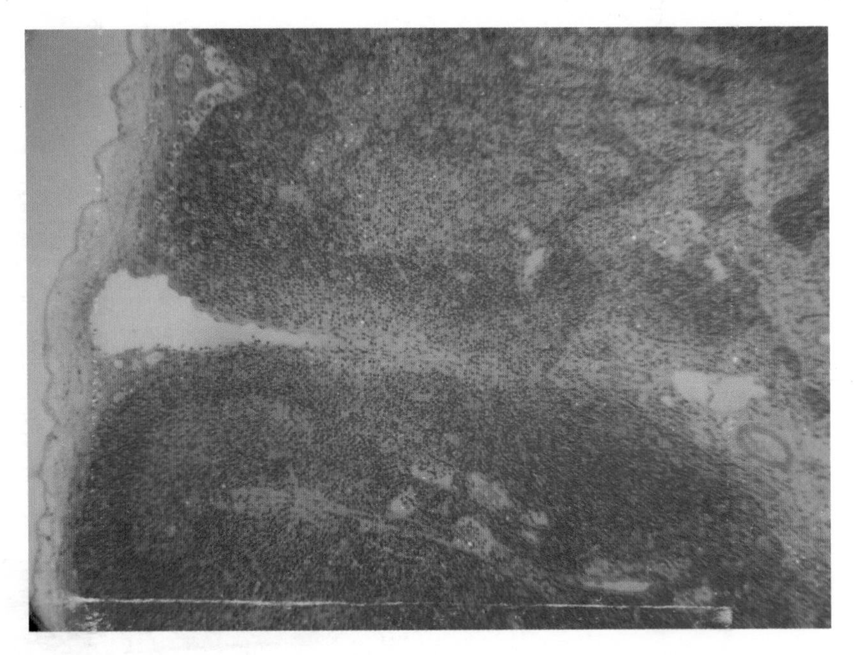

COLOR PLATE V. See Chapter 2, page 34, Figure 2.16a.